Developments in Aquaculture and Fisheries Science, 4

FARMING MARINE FISHES AND SHRIMPS

DEVELOPMENTS IN AQUACULTURE AND FISHERIES SCIENCE

1. FARMING MARINE ORGANISMS LOW IN THE FOOD CHAIN
   A Multidisciplinary Approach to Edible Seaweed,
   Mussel and Clam Production
   by P. KORRINGA
   1976 xvi + 264 pages

2. FARMING CUPPED OYSTERS OF THE GENUS *CRASSOSTREA*
   A Multidisciplinary Treatise
   by P. KORRINGA
   1976 x + 224 pages

3. FARMING THE FLAT OYSTERS OF THE GENUS *OSTREA*
   A Multidisciplinary Treatise
   by P. KORRINGA
   1976 xiv + 238 pages

4. FARMING MARINE FISHES AND SHRIMPS
   A Multidisciplinary Treatise
   by P. KORRINGA
   1976 xii + 208 pages

5. SOUND RECEPTION IN FISH
   edited by A. SCHUIJF *and* A.D. HAWKINS
   1976 about 200 pages

Developments in Aquaculture and Fisheries Science, 4

FARMING MARINE FISHES AND SHRIMPS

A Multidisciplinary Treatise

P. KORRINGA
Rijksinstituut voor Visserijonderzoek, IJmuiden (The Netherlands)

ELSEVIER SCIENTIFIC PUBLISHING COMPANY
Amsterdam – Oxford – New York 1976

ELSEVIER SCIENTIFIC PUBLISHING COMPANY
335 Jan van Galenstraat
P.O. Box 211, Amsterdam, The Netherlands

Distributors for the United States and Canada:

ELSEVIER/NORTH-HOLLAND INC.
52, Vanderbilt Avenue
New York, N.Y. 10017

ISBN: 0-444-41335-9

Printed in The Netherlands

PREFACE

If one broaches the topic "farming the sea" it is not only the industrial production of molluscan shellfish one has in mind, however important that may be. One rather thinks in terms of switching over from reaping the harvest from the sea, usually consisting of a variety of species, of which only few fetch a high price on the market, to growing a crop of valuable fish or shrimps under controlled conditions and in monoculture: in short, animal husbandry under water. By preference one should grow the fish in dense cultures, as a bio-industry, comparable with modern-style farming of chickens and pigs.

It is possible to grow fish, at least certain species of fish, in rather dense culture, and to produce an impressive quantity of valuable animal protein on a rather modest acreage. The greatest efficiency can be expected when one succeeds in raising a species of high market value to consumer size within one single season. This can only be expected under tropical or subtropical conditions, where one is not hampered by a prolonged growth stagnation in winter.

From an economic point of view, it is not only the overhead expenses for the construction and maintenance of ponds or enclosures, for personnel and electricity, one has to take into account, but also the costs for food — often the decisive factor. If one knows the price one can expect to make for the end product, and takes into account all the overhead expenses plus a reasonable figure for the "normal" mortality rate, one can figure out how much one can afford for food during the entire growth period. If one has reliable information on the conversion rate for a given food item to fish flesh, one can decide how much one could pay at the utmost for that food. Here one often finds constraints which make it economically unattractive to farm a given species of fish. If one wishes to farm a fish which is also brought to the market by fishermen, one can hardly hope to make a higher price for the farmed product than the fisherman gets for the same species. From this alone, one can easily see why farming cannot possibly be remunerative for the majority of fish species.

The best chances exist where it concerns a valuable fish which grows fast and can thrive on cheap food. These are in the first place the herbivorous fishes, living on plants big and small, or remains thereof. It takes only some manuring to stimulate the growth of such plants and hence the food bill can be kept low. Therefore farming of herbivorous marine fishes represents a category of special interest, also for developing countries, since one can produce a fair tonnage of valuable animal protein on very cheap vegetable food.

Unfortunately there are only a few species of marine fish which fall under this category. Two types are farmed on a large scale: the milkfish (*Chanos chanos*) and the mullets (*Mugil* species). Therefore, considerable attention is paid to the farming of these herbivorous fishes, both under tropical and subtropical conditions, respectively in Indonesia, and in Israel and Italy. The farming of these herbivorous fishes should not be seen as an industry to replace the ancient free fishery, for milkfish and mullets are hardly caught at all by fleets operating on the high seas. It is therefore an industry which supplements the existing fishery, and does not compete with it.

The carnivorous fishes and shrimps (which also have carnivorous habits) form quite a different category. For good growth they need a fair amount of animal protein in their food, and animal protein is almost everywhere quite expensive, even if one thinks in terms of fish meal. Experience has shown that some valuable species of carnivorous fish are quite fussy in their food preference and require high quality food, often fresh or deep frozen fish, to grow fast and to develop the special flavour appreciated by the consumer. Therefore farming of carnivorous fish has much in common with farming of chickens in a modern bio-industry. One needs a certain acreage of fish ponds in a good climate for rapid growth to marketable size. One can bring a lot of fish together in one single pond, but then good flushing and aeration becomes a necessity. The carnivorous fishes will never grow on the small amount of animal food produced in the pond on basis of the carbon dioxide assimilation by the green plants, big and small. One has to bring in food from a far greater acreage than the pond itself. If this food is trash fish, not suitable for human consumption, but caught in fair quantities together with more valuable fish species, one has a good chance to make the farming remunerative. If one has to buy deep frozen trash fish or high grade fish meal for this purpose, things may be more difficult. Pelleted feed may be used to farm carnivorous fish, but it is often the required dose of animal protein in the pellets which leads to a high, often too high, price for the food.

Farming carnivorous fish on a diet of species of fish which are also suitable for direct human consumption (herring, pilchard, mackerel, horse-mackerel, saury) is practised in some countries. One should realize that this means conversion of a large tonnage of low grade fish into a much smaller tonnage of high grade fish. Only rich countries, and those which can export their end-product for a high price, can afford such a procedure. It should be emphasized that such a type of sea farming should not be considered as an industry producing food to appease the hunger of the millions whose diet is too poor in animal protein, since it leads in reality to a reduction of the tonnage of edible animal protein on the market.

The same holds good for intensive shrimp farming. Where one practises extensive farming, at least partly based on natural food, and where one can find or compose rather cheap food, not withdrawn from the market of human food, to administer to the shrimps as supplementary feed, shrimp farming can become an interesting industry, especially for developing countries. Where one feeds the shrimps, however, a high grade food, one makes a luxury product at a high price, which is only acceptable for rich countries which are not short in animal protein.

To facilitate comparison with the cases of mussel farming and oyster farming I have decided to adopt the same subdivisions for each chapter as have been used in the three previous volumes, dealing respectively with mussel farming, with the farming of cupped oysters, and with the farming of flat oysters. Again, the section dealing with all the work involved in farming forms the bulk of the description:

A. BACKGROUND

A.1. General principles
A.2. Biology of the species
A.3. Geographical situation
A.4. Hydrographic pattern
A.5. Legal aspects

B. SEQUENCE OF OPERATIONS

B.1—Bx. This section forms the body of the descriptions, giving from the early developmental phases till the market, step by step, all the operations required, with indications of techniques used and of the amount of labour involved.

C. FARMING RISKS

C.1. Hydrographic conditions
C.2. Predators
C.3. Parasites and diseases
C.4. Competitors

D. ECONOMIC ASPECTS

D.1. Rents
D.2. Inventory items
D.3. Expendable items
D.4. Purchase of seed stock
D.5. Labour
D.6. Sales

E. GOVERNMENT ASSISTANCE

F. GENERAL IMPRESSION

G. BIBLIOGRAPHY

Where necessary, a section describing the technical details is added after the General Impression.

Again, the economic aspects are, as far as is feasible, rather expressed in quantitative units, such as the magnitude of the personnel, gallons of diesel oil used per year, tonnages sold, etc., than in monetary units. This will prevent the material presented in this book from becoming out of date. Still, some information is expressed in monetary units, and with the exception of Israel, where U.S. dollars are used (at the time of observation the U.S. dollar equalled 4.20 Israeli pounds), always in the money of the country concerned. A conversion table may serve to compare the local units with the American dollar. The figures are given for November 6th, 1972, a date representing more or less the middle of the period in which the data were collected, but to show how rapidly these figures may change a list of conversion rates for a more recent date (viz. March 20th, 1975) is added.

Currency	Abbreviation	Conversion rate 6-11-1972	Conversion rate 20-3-1975
American Dollar	US $	1.0	1.0
Indonesian Rupiah	Rp.	404.125	415.19
Italian Lire	L.It.	584.5	626.51
Japanese Yen	Y	301.0	285.23
Norwegian Crown	N.Kr	6.625	5.39

P. Korringa

ACKNOWLEDGEMENTS

Again, I feel first of all impelled to express my sincere gratitude to the leaders and proprietors of the enterprises selected as examples of regional fish or shrimp farming industries for enabling me to make use of their experience of many years standing, and for giving me a truly detailed insight into the labour involved in farming valuable species of fish or shrimps, and into the hazards and economics of their enterprises. Their names, noted in the heading of each chapter, need not be repeated here.

Further, I thank my colleagues who so kindly provided me with additional scientific information on the local hydrographic and biological conditions. For the case of mullet farming in Israel I have to express my heartfelt gratitude to Mr. S. Sarig of the Laboratory for Research of Fish Diseases at Nir-David for the painstaking way he formulated replies on my long questionnaire. Mr. Sarig was so kind as to collect a lot of information of a biological, technical and economic nature on fish farming in Israel for me, and replied extensively in several additional letters. He further kindly consented to discuss these matters orally on the occasion of his participation in the 1972 EIFAC meeting held in Amsterdam. Further, Dr. Y. Chervinsky and Mr. O.H. Oren provided me with useful information.

For the case of mullet farming in Italy I feel I must express my sincere gratitude to Dr. G. Ravagnan, proprietor and director of the fish farming enterprise described, for receiving me so kindly and for the enthusiasm with which he provided me with a wealth of supplementary information in a series of long letters and in various interesting reports on his findings and experience.

For the case of farming milkfish in Indonesia it was my colleague Mr. Rustami Djajadiredjo, director of the Fisheries Institute at Bogor, who kindly provided me with detailed information on the biological, technical and economic aspects of the fish farm selected as the example. When I visited Bogor again, recently, Mr. Rustami was kind enough to examine my manuscript critically and to give me valuable advice for improvement on some points.

For the cases of farming fish and shrimps in Japan I am greatly indebted to Mr. P. Businger, Agricultural Attaché of the Royal Netherlands Embassy at Tokyo, for collecting information on-the-spot for me on the basis of extensive questionnaires. During a sojourn in The Netherlands, Mr. Businger found time to discuss some points with me orally. Apart from the written information, I received series of photos to elucidate the cases. There were made specially by Mr. Businger himself, and by his secretary Mr. Y. Akimoto. I am very grateful for their efforts which enabled me to obtain a clear insight into the enterprises selected as examples.

For the case of farming of Salmonids in a Norwegian fjord I wish to express my gratitude to Mr. D. Møller of the Fisheries Institute in Bergen for accompanying me to the fish farm selected for the purpose, and for giving me valuable additional information. Further, I want to thank Dr. H.G. Gade of the Geophysical Institute of the Bergen University for providing me with interesting hydrographic information.

Again, I wish to express my sincere gratitude to my assiduous secretary, Miss E.L. van der Weele, for the painstaking accuracy with which she ordered my data and for her assistance in preparing the manuscript, and to Mr. B.L. Verboom for making charts and drawings to elucidate the various chapters, often working from difficult base material.

P.K.

CONTENTS

INTRODUCTION

Herbivores promise an interesting production level because of their short food chain: they feed directly on the organisms of the primary production. Unfortunately there are but few herbivorous species of fish in the marine environment suited to farming. The most important of these is no doubt the milkfish (*Chanos chanos*), which reproduces in offshore, high salinity water, but can be raised in water of an extremely wide range of salinities. Ponds constructed in the mangrove swamps fringing the coasts of several tropical countries can be used to advantage for milkfish farming. But it is also possible to raise this fish further inland, in water of low salinity, provided its favourite food — blue-green algae — is amply available. One glance at maps of countries like Indonesia and the Philippines shows that it must be possible to expand the milkfish industry considerably, for there are vast stretches of mangrove swamps which are at present not used at all, but which would be suitable for fish farming. There are, however, some constraints which hamper rapid expansion of this industry. Of course, there is the need for skilled farmers, but the present training centres can see to this, if required, at an accelerated pace. Further, one should find markets of reasonable size in the neighbourhood, markets which can absorb the farmed fish at a good price, for milkfish is by preference bought in the fresh state, though palatable preserves can be made of it. The main bottle-neck is, however, that it is often difficult to collect or to buy all the fry one needs for stocking the ponds. In several areas it becomes increasingly difficult to purchase fry in the desired quantities. One now explores fry catching areas further away, and even uses planes to transport the fry. The reasons for this scarcity are imperfectly known. Expansion of milkfish farming is to be expected from the moment it is learned how to breed this fish under controlled conditions, but that moment seems still quite far away.

Milkfish farming is not only interesting for the amount of animal protein it brings on the market, and that in no way in competition with the existing fishing industry, for milkfish is hardly caught at all in the open sea, but also to provide a living for the farmers, operating on family basis. The experience is that it is economically sound to bring the milkfish to the market long before it is fully grown. One can use the ponds most efficiently by keeping the fish in them for one season only. The conversion rate from blue-green algae to fish flesh is better in the younger fish than in the older ones. In some countries one not only adds fertilizer to the ponds to stimulate the development of blue-green algae, but also administers extra food, mostly in the form of pellets, to the fish. Thus the production per acre can be increased, but this not on the basis of the primary production in the pond itself. One comes then, close to a real bio-industry, in which the fertility and productivity of the site of farming no longer plays an important part. Production figures for such a situation must therefore be considered with caution, since the economy of such a system is quite different from that of farms based on local primary production.

A second herbivorous fish is the mullet, several species of the genus *Mugil*, which is farmed in brackish water ponds in several subtropical countries, predominantly in Israel and Italy. Mullet farming has much in common with milkfish farming, and encounters the same constraint: scarcity of fry of the best species, and the need to develop a system of reproduction under controlled conditions. The scarcity and expense of the fry makes it necessary to protect the juvenile fishes in indoor breeding houses and to transfer them step by step to outdoor ponds and to larger ponds for extensive farming. Just like milkfish farming, mullet farming does not compete with fishing, for the fishery does not normally land any quantities of mullets caught in the open sea. Even if mullet farming would in due course expand, it could not take the place of the present fishery. If, one day, fry were amply available through controlled breeding, one could even consider raising mullets in the temperate zone, on sites where cooling water from a power plant could be used to avoid too low water temperatures in winter, for low temperatures not only mean growth retardation, but even lead to the death of mullets.

Farming of carnivorous species is based on quite a different principle than the farming of herbivores. The ponds or netting enclosures serve only to prevent escape of the fish. For reasons of efficiency, one should stock the ponds or pens quite densely, which means, on the other hand, that flushing and aeration should be perfect, and that parasites and diseases have a better chance to find a victim than in more extensive fish populations. Feed is an important item in the budget of the farmer of carnivorous species. Therefore one will always select species for farming which bring high prices per kilogram on the market: species such as eel, salmon, shrimp, sole, turbot, lobster. When the tonnages landed by the fishery are limited, and in reality insufficient for the demand on the market, there is good reason to consider whether the species could be farmed. Just like in animal husbandry on the land, it is wise to grow the species in monoculture and to keep ponds and pens free from competitors. If it has been shown that it is, in practice, possible to farm a given species under controlled conditions, one should carefully study the possibility of a regular supply of young fry before constructing an expensive plant. By preference one should select a species which can be grown in full cycle from egg to egg, like chickens, thus to be independent of the whims of nature. Such cases do exist (e.g. farming salmon and rainbow trout in sea water), but in many cases one is still dependent on the catch of fry in nature. This leads not only to fluctuations in the annual supply, caused by natural conditions, but also, often, to mistrust from the side of the fishermen, who may fear a reduction in the stock of adults they fish by a too intensive fishing of young fry on behalf of the fish farmer. A clash of interests is certainly possible here. The farmer will claim that he protects the young fishes against predators and other dangers, so that ultimately, a greater tonnage will be brought to the market than would ever be caught by the free fishery; but on the other hand, the fisherman may claim that so few specimens might reach the adult stage in nature after intensive fishing for juveniles that reproduction may fail. This is the famous topic called "stock-recruitment relations".

The item food weighs very heavily on the farming of carnivorous species, because one dares not deviate too much from the natural food. Therefore one feeds the fish with fish of other species. As a rule such feed fish should be fresh or deep frozen, and often one buys fish which could be good for the normal consumption market. From the point of view of national economy, this type of farming may offer a good living to the farmers, often operating on family basis in remote areas, but it produces a rather small tonnage of high-priced luxury food at the cost of a large tonnage of low-grade consumption fish. Therefore efforts to develop pelleted feed from cheap ingredients should be encouraged, provided the fish reared on such pellets can meet the high standards of flavour for which they are renowned. Until this phase has been reached one cannot claim that farming carnivorous fish and crustaceans will lead to greater tonnages of good animal protein appearing on the market, or that this type of farming will replace the ancient fishing in due course. If an ingredient like fish meal were required in fair percentages in the pellets one is trying to develop now, it is the production of fish meal which would in due course limit the potential production of farmed fish. In that case, a total production of 10 million metric tons of fish is already unrealistic, for the quantities of fish meal required for such a production simply cannot be produced and bought in competition with chicken and pig farmers.

Chapter 1

MULLET FARMING IN ISRAEL

Example
Fish farm of the "Ein-Hamifratz" kibbutz, 25210, Israel.
Period of observation: 1971—1972.
Informants: D. Mires of the Ein-Hamifratz kibbutz (technical and economical data); S. Sarig of the Laboratory for Research of Fish Diseases at Nir-David, 19150, Israel (biological data).

A. BACKGROUND

A.1. General principles

This is pond farming of mullet under sub-tropical conditions. Mullet are marine fish with katadromic tendencies. At a given time of the year, their young fry congregate in the estuaries of the Israelian rivers, and can be caught there by hand-nets; to be transferred to fish ponds, where they are reared in a mixed culture in which carp predominate. The ponds contain slightly brackish water and receive a fair dose of fertilizer; the carps are fed daily with sorghum and with pellets rich in protein. This indirectly enhances the feeding conditions for the mullet, which reach marketable size in their second summer.

Farming in densely stocked ponds involves attacks of parasites and diseases. Government laboratories give advice on their control. Despite intensive research, it is not yet possible to produce mullet fry on commercial scale under controlled conditions, so the fry have to be collected in the wild. Hence one should speak of a semi-culture. Of the various species of mullets entering the Israeli rivers, only two can be used for farming: *Mugil cephalus* and *Mugil capito*.

A.2. Biology of the species

Mullet belong to the family of the *Mugilidae*: marine fish of slender shape, almost circular in cross-section, and covered with round scales. They have a flat, snub snout, a small beak and show four spines in the first dorsal fin and one or two in the second dorsal fin. They live chiefly on vegetable matter of microscopic dimensions, dead and alive, and also partly on small zooplankton organisms. They collect micro-organisms and detritus from the bottom, and from places with a lush vegetation, such as pilings. They chew and select the sucked-up material carefully, spitting out indigestible objects.

The spines of the gill-arches form, together with the oesophagal teeth, a finely meshed sieving apparatus. The strong and muscular stomach, covered inside with a horny layer, as for birds, serves to grind the ingested food. The intestinal tract is unusually long for a fish, about seven times its body length.

The mullet are mainly found in tropical and sub-tropical waters, principally in the inshore waters. They are very euryhalinic and often migrate into estuaries and lagoons. Spawning always takes place in the high sea, but very little is known on the exact whereabouts. Investigations carried out at sea revealed that the smallest fry is to be found well offshore. When the fry is about 20 mm long, it enters the estuaries in search of good feeding grounds. The larger mullet are very lively fish, and difficult to catch since they do not easily take bait, which is logical in the light of their feeding habits. When approached by a net they jump high out of the water over the top of the net, or dive into the mud underneath its ground-rope.

Not all six species of mullets occurring at the Mediterranean coast of Israel lend themselves to fish farming. Some species show poor growth and survival, while others occur in insufficient numbers. The best are *Mugil cephalus* (L.), which shows the best growth of all, and *Mugil capito* Cuv.

A.3. Geographical situation

The Ein-Hamifratz kibbutz is situated on the northern part of the coast of Israel, about 13 km north-northeast of Haifa. Its exact geographical position is 32° 54′N and 35° 05′E, close to the mouth of the river Naaman, precisely

Fig. 1. *Mugil cephalus* of marketable size. (Photo Archives Laboratory for Research of Fish Diseases, Nir-David, Israel.)

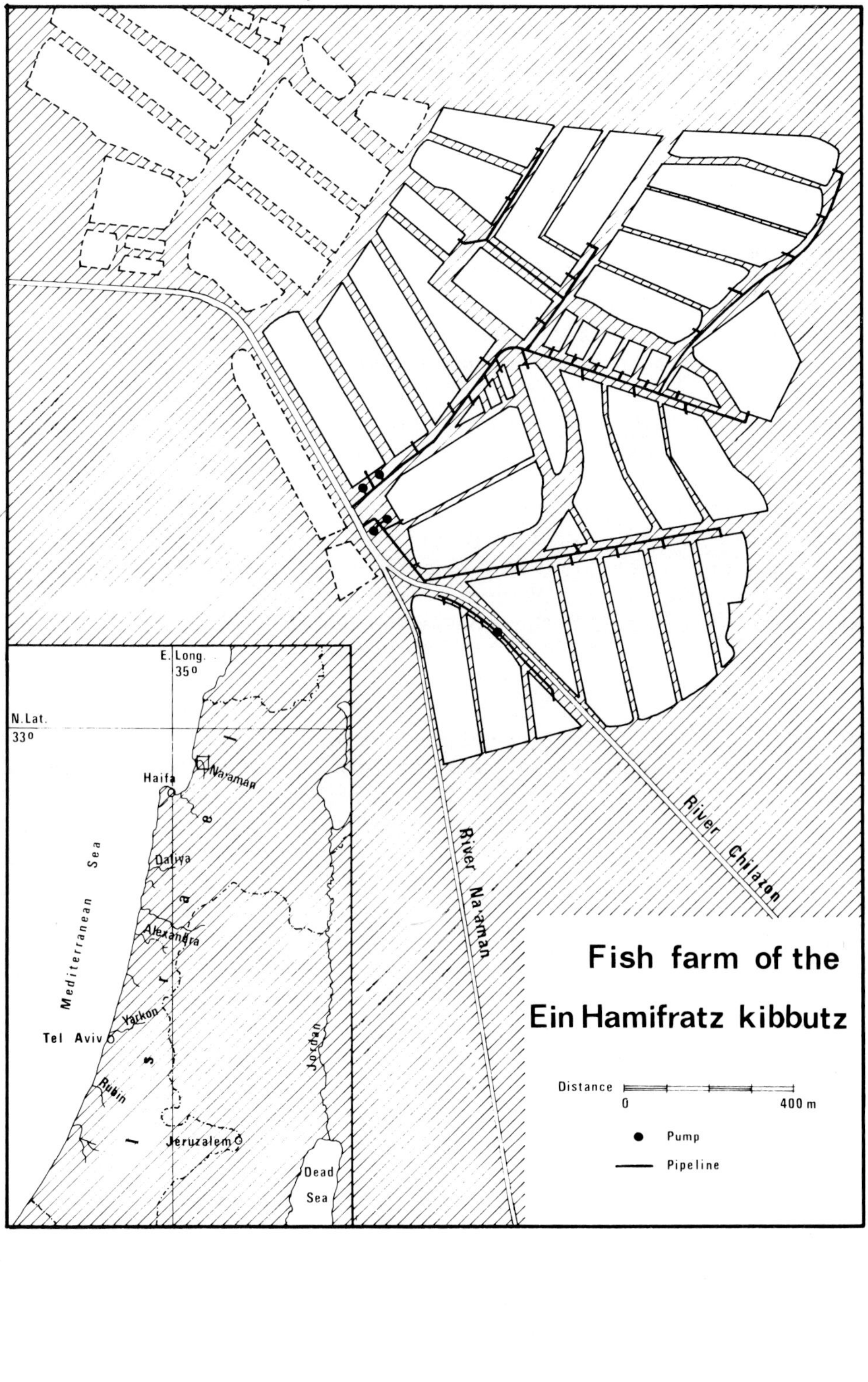
Fish farm of the
Ein Hamifratz kibbutz
Distance
0
400 m
Pump
Pipeline
River Na'aman
River Chilazon
E. Long.
35°
N. Lat.
33°
Haifa
Na'aman
Daliya
Alexandra
Yarkon
Tel Aviv
Rubin
Jeruzalem
Jordan
Dead Sea
Mediterranean Sea

where its tributary the river Chilazon debouches, about 2 km from the sea. The Mediterranean sea-shore has there, a sandy beach, sloping gently down to a depth of 6 to 7 m, some 100 m from the water-line.

The farm consists of 130 ha of ponds, 36 in total, mostly 3.5 to 6 ha, but with some small breeding and storage ponds. The ponds were dug in 1947. The fish farm belongs to the Ein-Hamifratz kibbutz, which is west-southwest of the fish farm.

A.4. Hydrographic pattern

Shape and depths of the ponds

The ponds of the Ein-Hamifratz fish farm are grouped more or less fan-shaped around a bend in the downriver section of the river Naaman. Most of the 36 ponds are rectangular in shape, with a width of about 100 m. The ponds have a muddy bottom and a water depth of 1.5 to 2 m. The banks slope gently, at about 1 : 4, and are coated with stones; at higher levels with grass. To control the water level of each individual pond a concrete structure equipped with slides and screens, popularly called "the monk", has been erected at the deepest part of the pond, called "the fishing pit". It takes 2 to 3 days to drain a large pond of 5 to 6 ha, and of course, a shorter time for the smaller ponds. Pumps are used to fill the ponds and to keep the water level at the desired height when there is loss due to evaporation and seepage.

Salinity

A very important feature is the salinity of the water to be used for the ponds. Fresh water being a rare and precious commodity in a semi-arid country like Israel, the Government does not allow its use for fish ponds. Fresh water with a chlorinity lower than 275 mg/l (i.e. 0.5‰ salt) should be used for drinking water and for irrigation only. Fish farms are constructed where th river water and the ground water show a chlorinity of over 350 mg/l. This is very often the case, since salt and brackish springs and wells abound in many places. The water used to fill the ponds of the Ein-Hamifratz kibbutz has a chlorinity of about 800 mg/l (1.5‰), but in the summer season evaporation may bring it up to 2 000 mg/l (i.e. 3.7‰). The water of the Mediterranean sea has here a salinity of about 38‰. Mullets, being very euryhaline, can thrive in a wide range of salinities, but carp, the main crop of all Israelian fish ponds, get into trouble when the chlorinity rises above 2 500 mg/l (4.5‰). The water is pumped from the Naaman river, but when the river level is very low, the water in the ponds must be used as economically as possible: in emptying a pond, the water is then not discharged into the river but pumped into another pond.

Water temperature

The water temperature in the ponds varies with the season. The winter minimum may be as low as 8° C, the summer maximum as high as 29° C. The optimum temperature for growth of the fish is in the range of 20—28° C, which is obtained from April to October. In the surface layers, the daily fluctuation may be 3 or 4° C, in the deeper layers only 1° C. *Tilapia* get into trouble at water temperatures under 10° C, and all farmed fish species are seriously endangered by temperatures over 32° C.

Oxygen and pH

Since the ponds are enriched, which leads to a dense growth of plants of various description, mainly of microscopic size, a far greater daily range in the level of oxygen and pH is to be expected than in the open sea. This is in fact the case: in summer from 15 to 20 mg/l O_2 has been recorded in the middle of the day in the surface layers, and 8 to 12 mg/l near the bottom. Around sunrise the figures obtained were 4 to 5 mg at the surface, and about 2 mg near the bottom. Occasionally, especially during a water bloom, more extreme values can be found: up to 28 mg in the middle of the day, and down to 0 mg just before sunrise. Such a situation may lead to fish mortality. To avoid very low oxygen levels, pumping in of river water is always done at night. In order to keep the pond in good condition 150 to 200 m^3/ha (i.e. about 60 000 m^3/ha/year) of water should be pumped in.

The pH shows also a considerable daily fluctuation parallel with photosynthetic activities. In some of the ponds of the Ein-Hamifratz farm, a pH of 9.4 was recorded at noon, and only 7.7 in the second part of the night. The pH range of 7.0 to 9.5 should not be inconvenient for the fish species farmed in the Israelian ponds.

Plankton and other vegetation

In the section on farming risks it will be noted that noxious types of vegetation, toxic or those leading to tainting of the fish, may occur, and should be controlled with the greatest care. Generally the water in the ponds is very green, a white object becoming invisible at a depth of about 20 cm.

A.5. Legal aspects

The land on which the fish ponds are situated belongs to the State. The Ein-Hamifratz kibbutz has rented it for a period of 49 years for a lease of U.S.$ 4/ha (1 ha = 10 "dunams").

Israel has 83 fish farms, covering a total acreage of 4 800 ha; 98% of those are organized as kibbutzes. The Ein-Hamifratz kibbutz totals 160 families. Its main concern is agriculture (wheat, barley, maize, sorghum), horticulture (grapefruit) and growing chickens and turkeys, but also some industry (packing material). The fish farm is also one of its activities, employing eight men full-time.

A kibbutz is directed by a board of about 9 functionaries, among whom are a treasurer, an economist and someone organizing the labour-pool. This board is led by a secretary. They all are elected for a limited number of years. Of the total budget of the Ein-Hamifratz kibbutz, a certain proportion is earmarked for the fish farm; both for its running expenses and for buying inventory items, all of which are carefully estimated. Unforeseen expenses due to emergencies require a detailed explanation before granting the money will be considered. Profit made by a section goes to the central treasury of the kibbutz. No salaries are paid, but every member is provided with the necessities of life, on a personal budget broken down into various items, including such things as shoes and books, not to be interchanged. Eating at the central restaurant is free, and so is education and smoking. A small amount of pocket-money is paid and some funds are set aside for travelling. When the economic situation of the kibbutz turns unfavourable, items such as travelling and investment are cut first of all, clothing and feeding last.

The fish farm team elects one man as their leader, but this does not, however, affect his financial position. This man in charge of the fish farm may serve for several consecutive years if he does a good job. He has to work at the fish farm with his seven co-workers, but in case of a heavy seasonal workload he can apply for extra hands from the labour-pool. If carefully planned, this may lead to great efficiency in labour and avoidance of idling.

B. SEQUENCE OF OPERATIONS

B.1. Procuring the fry

Despite serious efforts in many countries, a practical system of maturing and spawning mullets under controlled conditions has not yet been developed. The occasional success has been achieved on a laboratory scale, but thus far, there is no practical system to produce the quantities of fry needed for farming. Still, there is every hope that this will eventually be the case. For the time being, the mullet fry to be stocked in the ponds have to be collected in the wild.

Mullet spawn somewhere offshore in the sea, but the exact position where the mullet of the Israeli coastal waters spawn is not known. The young fry arrive at the coast about 18 to 22 mm in length, and try to enter the estuaries. The Israeli rivers do not discharge very much water, and in the dry season, the run-off is usually even so low that the sea builds a bar of sand across their mouths. Western gales, which occur frequently in November, break the mouths open again and then the mullet fry enter the estuaries. The ports of entrance are, from north to south, the estuaries of the rivers Naaman, Alexandra, Yarkon, Rubin and Aza, of which the first, the third and the fourth are the more important, the others being too insignificant or too polluted. The fry will not enter an estuary when its water is grossly polluted; therefore it is of paramount importance for fish farming to keep the rivers as clean as possible.

Six different species of mullet may be found in the estuaries of the Mediterranean coast of Israel, but the farmers are only interested in the fry of *Mugil cephalus*, entering from late October till February, and in that of *Mugil capito*, which makes its appearance from January to April. If the winter storms come late, fry of the valuable *Mugil cephalus* will be hard to find.

Each of the 65 farms raising mullet holds a team of three to six workers ready to collect fry at the time they appear. They travel by car from estuary to estuary looking out for mullet fry. They wade out in the estuary fishing with dipnets [1*] and hand-seines [2] made of mosquito netting (Fig. 2). Boats are not used for collecting mullet fry. The fishermen carry containers with them, which are emptied from time to time into the 1 m^3 tank installed on their van. The water in this tank is duly oxygenated, for which the oxygen is carried in metal cylinders.

Fig. 2. Collecting *Mugil* fry in the estuary of the Naaman river with a fine-meshed hand-seine. (Photo Archives Laboratory for Research of Fish Diseases, Nir-David, Israel.)

---

* These reference numbers refer to a more detailed description in the section "TECHNICAL ASPECTS" at the end of the chapter.

There is fierce competition in this field, since all farmers are anxious to get their supply of fry, and since it is not always easy to catch the numbers needed. Often the fishermen are working day and night in the estuaries. Per hectare of fish pond in operation a farm will need about 3 000 young mullet. Catches of 3 000 to 10 000 fishes per day are considered normal during the peak of the run, but conditions are not always ideal and there is an undeniable downward trend in the total catches of mullet fry, which fluctuate as widely as from one to 10 million individuals per year. When a farmer cannot catch his share he is forced to buy fry from individual fishermen at a price of US $ 22 per 1 000, delivered at the farm. Fish-farmers are worried about the decline in the catches of fry. They wonder what may be held responsible for this, and claim it might be the construction of the Asswan dam in the Nile river, decreasing the fertility of the sea. They do not believe that they are to blame themselves, giving so few mullet a chance to join in spawning in their natural habitat by transferring them to their ponds from which they will never return. They hope that the decline is to be ascribed to natural fluctuations in reproduction, but will only feel at ease when a good system of controlled reproduction has been developed. If not, it might be worthwhile to consider allowing some of the mullet kept in ponds to return to the sea, to breed a progeny which will, in due course, strengthen the numbers running in the estuaries.

Experienced fry fishers are able to recognize whether they have caught the correct species of fish, but at arrival on the farm, somewhere in Israel, occasionally even 200 km inshore, the leader of the fish farming team will take his microscope to check what percentage of the catch belongs to the desired species *Mugil cephalus* and *Mugil capito.* Differences in pigmentation and in the pyloric caeca can be used to identify them (Perlmutter, Bograd and Pruginin, 1957; Yashouv and Berner-Samsonov, 1970). If the catch happens to consist of over 70% of fry of *Mugil auratus* or other unsuitable species, the leader may decide to discard the whole lot.

Each tank arriving at the farm may contain up to 10 000 small mullet, but there is usually 30% loss due to damage during capture and transport.

B.2. Stocking the ponds

The mullet fry is stocked at a density of 10 000 to 20 000/ha in the nursery ponds, which are of the same pattern and depth as the fattening ponds, but only smaller in size, viz. from 2 to 4 ha. Small tanks mounted on barrows with pneumatic tyres, and provided with oxygen, for which the steel cylinders are carried on a second barrow, may serve to bring the fry to the various nursery ponds. Despite all precautions taken, there is loss of 30% of the fry between the estuary and the nursery pond.

Once in the nursery pond, the young mullets keep company with the carp, some 1 500/ha, and with some *Tilapia.* These fish are, however, noticeably larger when put in the nursery ponds, the carp 250—350 g each, the *Tilapia* 40—80 g. The purpose of putting in the large carp is to prevent development of a dense vegetation of the stonewort *Chara*, and of filamentous algae such as *Spirogyra*, on the bottom of the ponds. The carps are constantly rooting up the bottom in search of Chironomid larvae and other food items, hence the unwanted plants do not get a chance to spread. Carp and *Tilapia* do not feed on mullet fry. *Tilapia* is put in to feed on any surplus phytoplankton and to increase the yield of the pond. This holds for the carp too.

The nursery ponds are duly fertilized: 200 kg/ha of superphosphate, given in one dose in spring, and 300 kg/ha of ammonium sulphate, given in weekly applications throughout the summer. This promotes development of phytoplankton and zooplankton, and also leads to a richer bottom fauna. The carp in the nursery ponds should also get some supplementary food in the form of sorghum and special pellets containing some 25% of protein [3]. The quantities to be administered are determined with the aid of feeding tables produced by the Dor Fish Culture Research Station. In principle, the feeding rate is 5% of the body weight of the carp per day, i.e. 20—30 kg/ha/day.

The young mullet grow rapidly in the nursery ponds; in April they reach a weight of 15 g, and towards the end of their first summer the *Mugil cephalus* weigh 70—100 g and *Mugil capito* 40—50 g. They thrive predominantly on green planktonic algae, such as *Scenedesmus* and *Pediastrum*, but also take some small zooplankton organisms and suck up bottom detritus in quantities. When growing up, they may also take some of the pellets meant for carps, but by far the largest proportion of their food consists of the natural items mentioned above, the development of which has been promoted by the fertilizers.

B.3. Transfer to fattening ponds

In winter or early spring, the mullet are transferred to the 4 to 6 ha fattening ponds. The nursery ponds are drained completely, which makes the fingerlings congregate in a section of the pond called "fishing pit", this being about 80 cm deeper than the remainder of the pond. When the water is low enough, which takes about one day, men wade out in the pond and use seine nets to concentrate the fish still further; steel supports can be put under the headline of the net to keep it well above the surface of the water, for the mullets will try to escape by jumping high into the air. Next the fish elevator [4] is used. The fish are scooped up with large perforated buckets to be dumped in the 70 l perforated container mounted on the elevator (Fig. 3). The container is lifted mechanically over the rim of a large rectangular tank, mounted on a truck waiting on the embankment, into which it is emptied. The water in the tank is thoroughly aerated. In this tank the fingerlings are transported to the fattening ponds. Now that the fishes are larger and

Fig. 3. The fish elevator in operation. The fish are concentrated in the net and the container dumps its cargo into the tank mounted on a truck. (Photo Archives Laboratory for Research of Fish Diseases, Nir-David, Israel.)

stronger, losses due to handling are smaller, from 5 to 10%, considerably less than in handling the fry.

When the balance is calculated, after taking the fish from the nursery ponds, it will be seen that the mullet have suffered some losses during their sojourn in these ponds; as a rule from 10 to 50%. These losses are inflicted by various agents: delayed effects of injury during the stocking procedure, lack of oxygen, toxic plankton, parasites, and to a lesser degree predators.

The stocking programme is as follows: per hectare 2 500 carps of 200 g, 1 500 *Tilapia* of 50 g and 1 000 mullets of 50 to 100 g. Sometimes silver carps, *Hypophthalmichthys molitrix*, originating from the Amur district in northeastern Asia, are also added; these use very small planktonic algae as their main food in the ponds, hence they hardly compete for food with the other species. The fattening ponds are fertilized in the same way as the

Fig. 4. Small tank for transporting live fish. In the background is a large transportation tank with aerator. (Photo Archives Laboratory for Research of Fish Diseases, Nir-David, Israel.)

nursery ponds: 200 kg/ha of superphosphate in one dose and 300 kg/ha of ammonium sulphate, now in bi-weekly applications. In addition 1 000 kg/ha of chicken droppings are spread bi-weekly.

Again, the carp receive as supplementary food sorghum and/or pellets, for which one consults the feeding tables. These take into account the weight of the individual carp, the total tonnage of carp/ha and the water temperature; larger amounts of food are given in periods of higher water temperatures. On average, 40—60 kg/ha/day are required for a stocking as described above. The food is administered once a day with a blower mounted on a tractor. This drops the food on a special place in the pond called the "feeding table", some 20 m offshore. The carp come into action as soon as they hear the engine of the tractor.

The mullets remain the whole summer in the fattening ponds, where they grow to the marketable weight of 500 to 800 g in 120 to 170 days. The average individual weight increment is 3 to 4 g/day. They take advantage of the fertilization of the pond, and only to a very minor degree, of the food administered directly.

B.4. Harvesting and shipping the marketable fish

To harvest the fish, the fattening pond is drained completely, which takes 2 to 3 days for a larger pond. The fish congregate again in the deeper fishing pit and are caught after further concentration with a seine net. Of course the same number of fish are not present as were originally introduced in the pond, for it is hard to avoid predation completely. As a rule the losses in their second summer vary between 5 and 15%. The fish are sorted, and specimens considered too small will be put back in a fattening pond for one more season. The marketable specimens of *Mugil cephalus* weigh 400 g, of *Mugil capito* 250 g. These weights do not indicate termination of growth; in ponds in the Lake Tiberia area *Mugil cephalus* has been grown, in a period of 4 to 5 years, to a weight of about 10 kg. The mullet are thoroughly washed on stainless steel washing tables fitted with douches to eliminate all silt, are neatly packed in plastic boxes (complete with head and entrails) and are

Fig. 5. Sorting and packing mullets for the market. (Photo Archives Laboratory for Research of Fish Diseases, Nir-David, Israel.)

Fig. 6. A box of mullets ready for the market. (Photo Archives Laboratory for Research of Fish Diseases, Nir-David, Israel.)

finally covered with crushed ice. They can be kept in cold storage overnight to be transported by a truck of the kibbutz to one of the 18 wholesale centres around the country, where they are sold by auction. The distribution is organized by quota via the Fish Breeders Association. All mullets are sold as fresh fish, none are used in canneries. The sales have no marked seasonal pattern, except that April and May usually show somewhat lower figures.

C. FARMING RISKS

It is not surprising that intensive fish farming in ponds, as carried out under the sub-tropical climatological conditions prevailing in Israel, involves a variety of risks, among which are adverse hydrographic and biological conditions, predators, parasites and diseases. Fortunately the Government provides research and advice in this field.

C.1. Hydrographic conditions

Salinity, temperature and pH remain within the range of tolerance for the fish species farmed here, but occasionally the amount of oxygen dissolved in the water of the ponds gives reason for concern. Since oxygen figures are lowest in the second part of the night, when consumption has exceeded production for many hours in succession, pumping in of river water is done at night. No special apparatus to mix the pondwater is used in the fish farm of the Ein-Hamifratz kibbutz.

C.2. Biological conditions

One of the greater risks is a bloom of the toxic Chrysomonad flagellate *Prymnesium parvum* Carter, a typical denizen of slightly brackish water. Its first record for Israel dates back to 1947, but now it is endemic in all brackish water ponds and constitutes a great potential danger to fish farming, some years leading to losses exceeding US $ 100 000. In about 3 to 5 days a bloom may spring up, reaching a density of 8×10^5 cells/ml. In that phase, fish will be affected by the poisonous character of the water, brought about by the toxin this flagellate excretes when its population has become dense. If no action is taken at that moment, a large scale mortality of all fish should be expected. Fish farmers should be on the look out for *Prymnesium* at frequent intervals, taking samples from their ponds for microscopic inspection at least once a week. The Nir-David Laboratory for Research of Fish Diseases gives advice on how to control this dangerous micro-organism. In the warmer season, 10 ppm ammonium sulphate reduces its population within 20 h to 50%, especially if administered during the pH peak at noon, but in winter one should take 2—3 ppm copper sulphate. From levels of 10 ppm upwards this is also dangerous for the fish. Liquid ammonia 10—15 ppm can also be used, but it requires special precautions and apparatus to spread this in the ponds.

Some species of blue-green algae may bloom in the ponds: *Microcystis*, *Anabaena*, *Oscillatoria* and *Spirulina*. Such blooms are facilitated by the high degree of fertilization of the water in the ponds. *Microcystis aeruginosa* produces a potent toxin; all species may at the end of a bloom take too much oxygen from the water for their breakdown, and some, especially *Oscillatoria*, may taint the fish so much as to render them unsalable. The "muddy" taste develops especially in ponds with a sandy bottom. Drifting algae, blown to a corner of the pond by the wind, can easily be destroyed by spraying 3% copper sulphate directly on it, taking care that the concentration in the pond does not exceed 1.5 ppm. A good method of control of such algae is spreading copper sulphate on the bottom of a drained pond, 100 g/m^2. This treatment prevents tainting of the fish flesh by excessive algal growth. After 3 days the pond can be filled again, 5 days later it can be stocked with fish. Such treatment helps for a period of 8 to 12 months. This expensive treatment is only applied in rather small storage ponds. Before stocking those again, repeated flushing is necessary to eliminate excess copper.

C.3. Predators

Various organisms try to feast on the dense stocks of fish in the ponds. Cormorants and several species of herons are to be feared among the birds. The former, being migratory birds, are reduced in number by shooting, the latter by destroying their eggs or young in the rookeries. Smaller nursery ponds may be covered with nylon netting to keep the birds away.

One should also keep an eye on the activities of the grass-snake *Natrix natrix*, and lay out special snake traps on the surface of the water when they are predating on the fish. Sometimes hundreds of them are caught with such traps.

C.4. Parasites and diseases

A variety of parasites may attack the mullet, and the crowded population in a fish pond greatly facilitates their spreading. Special mention should be made of the parasitic copepodes *Lernaea cyprinacea*, *Argulus pellucidus* and *Ergasilis* in several species. The latter parasite infests the gill filaments of the mullet and are sometimes present in hundreds (Fig. 7), leading to poor growth, emaciation and death of the host (Fig. 8). Further, trematodes, especially the digenetic trematode *Heterophyes heterophyes* may wreak havoc in the ponds. The latter species requires extra caution since it can

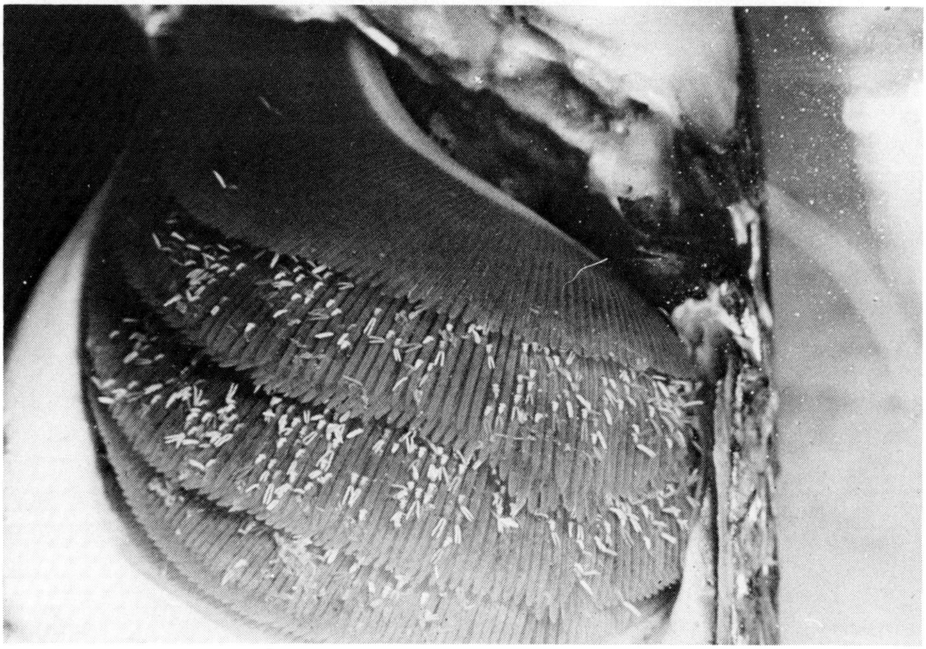

Fig. 7. Heavy infestation of the gills of a mullet with the parasitic copepod *Ergasilis sieboldi*. (Photo Archives Laboratory for Research of Fish Diseases, Nir-David, Israel.)

Fig. 8. Above, a normal mullet; below, an emaciated specimen of the same age, suffering badly from infestation with the parasitic copepod *Ergasilis sieboldi.* (Photo Archives Laboratory for Research of Fish Diseases, Nir-David, Israel.)

affect the human intestine if the fish is not properly cooked. *Myxosporidium* may also infest gills, fins and intestines of the mullets. The fungus *Saprolegnia* may be held responsible for killing mullet injured during transfer operations. Even slight injuries caused by loss of scales may lead to *Saprolegnia* infection.

It is clear that remunerative fish farming would be impossible here because of the presence of all these parasites, were it not that a specialized laboratory for research in this field had been established at Nir-David to advise and instruct the fish farmers of Israel. A great number of pesticides have been tested by this laboratory and the farmers are duly informed about their use, about the active concentration expressed in ppm, together with the safety index: the relation of the minimal dose killing the fish to that killing the parasite. The pesticides are applied to the fish in the ponds, only exceptionally (for carp) to fish taken out of the water. The leaders of the fish farming teams know how to use a microscope and can identify the more commonly occurring parasites themselves. Sarig's booklet gives an excellent outline.

D. ECONOMIC ASPECTS

All figures are given in US $ since the Israeli Pound is not officially convertible. There are about Is. £ 4.20 per US $.

D.1. Rents

As has been recorded above, the Ein-Hamifratz kibbutz rents 130 ha of ponds from the Government and has to pay an annual lease of US $ 4/ha.

D.2. Inventory items

The most important inventory items are the fish ponds themselves. The ponds of this kibbutz have been dug as long ago as 1947, when bulldozers and other heavy equipment was used. Since a 10% annual amortization was calculated, the construction costs of the ponds no longer weigh upon the budget of this kibbutz. One should count, however, on substantial repair of the ponds every 5 or 6 years. New ponds would now cost US $ 1 200 per ha (= per 10 dunams) to dig.

Important are the motorized pumps to pump up river water, and in summer to transfer water from the one pond to the other. The Ein-Hamifratz kibbutz has installed seven pumps, all near the river, producing 500 to 1 200 m^3/h water. The pumps require good maintenance and should be replaced periodically.

An immovable item is the shed for storing nets and other utensils. The list of chattels is rather long: the fish elevator; 3 tractors for various transportation needs, including the blower which distributes the food for the carps; tanks mounted on barrows and trucks for transport of fry from the estuary to the farm, and within the farm to the nursery ponds, the larger one for transfer of fingerlings in the farm, and of live carp to the market; steel cylinders for air and oxygen, mounted on trucks and barrows; some small plastic boats with outboard motors for work in the pond, and stainless steel washing and packing tables. An isothermic van to transport the mullet to the market does not feature on this list, being hired when necessary.

D.3. Materials

If they have to be bought, mullet fry cost at least US $ 22 per 1 000, delivered at the farm. A price as high as US $ 45 has been asked in 1972. Superphosphate costs US $ 50/metric ton and ammonium sulphate US $ 30/ metric ton. Chicken droppings are a by-product of the kibbutz itself. Sorghum costs US $ 50/metric ton and pellets US $ 100/metric ton. Pumping water costs about US $ 7/1 000 m^3 for electricity, and since a daily input of 150 to 200 m^3/ha is needed to compensate evaporation and seepage, a total of 60 000 m^3/ha/year has to be reckoned on. Various types of pesticides will cost US $ 80/ha/year. The plastic boxes for transport of marketable fish belong to the wholesalers.

From the description of the sequence of operations one can calculate how much money is required to pay for fertilizers, food, water and pesticides per hectare of pond per year, or, if one wishes so, for the whole life-cycle of a mullet per hectare.

D.4. Labour and personnel

It is impossible to specify how many men-days are devoted to mullet farming, and how many to farming of carp and *Tilapia*, in the Ein-Hamifratz kibbutz, since one works, as is usual in Israel, with a mixed culture, carp constituting about 75% by weight of the end-product. If counted separately the costs for catching fry come to about US $ 25 per 1 000. An important labour item is checking for *Prymnesium*, which takes the full time of one man during 9 months, to be calculated at a cost of US $ 12 per day.

It should be remembered, however, that the fish farming team of the Ein-Hamifratz kibbutz consists of seven full-time persons, while two to three extra hands are drawn from the central kibbutz in time of high work load, i.e. when a pond is to be emptied, and that a kibbutz does not pay salaries in money but in kind.

Since this description focusses on the production of mullet, being marine

Fig. 9. Fish elevator with long sloping rail, used to transfer fish from the ponds to the tank mounted on a truck. (Photo Archives Laboratory for Research of Fish Diseases, Nir-David, Israel.)

fish, no details are given of the first phase of producing carp and *Tilapia* in special breeding ponds. This procedure is the very same as in carp farming all over the world. The amount of labour involved is rather modest.

D.5. Sales

The Ein-Hamifratz kibbutz produces on an average about 20 metric tons of mullets per year. At the market, mullet of 500 to 800 g fetch US $ 1.40 to US $ 1.75/kg, mullet of 250 to 400 g, US $ 1.00 to US $ 1.30/kg. The prices fluctuate since mullet are sold by auction. For live carp, on the other hand, the price to be paid by the retailers is set by the Government at US $ 0.50/kg, for *Tilapia* US $ 0.65/kg. Carp make up some 70% by weight of the production of the Ein-Hamifratz fish farm. The remaining 30% consist partly of mullet, partly of *Tilapia.* The production of silver carp is still small.

Production of the Ein-Hamifratz farm (in metric tons)

Year	Carp	*Tilapia*	*Mugilidae*	Silver carp
1968	146	30	28	—
1969	135	45	16	—
1970	139	32	16	1
1971	156	41	21	10

D.6. Losses

As has been explained above, there are bound to be numerical losses in several phases of farming, even in a well-managed farm as that of the Ein-Hamifratz kibbutz: 30% on transfer from the estuary to the nursery pond, 10 to 50% during the sojourn in the nursery pond, 5 to 10% on transfer from the nursery pond to the fattening pond, and 5 to 15% during the sojourn in the fattening pond. Theft or poaching does not occur.

D.7. Further development

The production of farmed fish is still on the increase in Israel and the production per hectare of pond is remarkably high: 2 000 to 4 000 kg. It is, however, especially the carp which shows increasing figures. The difficulty of raising the production of mullets is found in the declining catches of fry in the estuaries. Until a good system of artificial breeding is developed for mullet fry, which would give the possibility of switching over from semi-farming to full-farming, the farmers will be totally dependent on the runs of fry which occur naturally in the Israeli estuaries, or perhaps on fry imported from estuaries in other countries, a practice not yet tried. There is reason for concern, since a downward trend in the numbers of fry caught is being noticed.

E. GOVERNMENT ASSISTANCE

The Israeli Government promotes fish farming in various ways. Of very great importance are the Fish Culture Research Station at Dor and the Laboratory for Research of Fish Diseases of Nir-David. The former gives advice to the farmers free of charge, i.e. advice on stocking, feeding and fertilization. The latter on control of parasites and diseases. The extension services do an excellent job and the farmers can keep pace with the results of research by reading "Bamidgeh", the "Bulletin of Fish Culture in Israel". Several scientists of the Hebrew University contribute to the knowledge of various aspects of the biology of the species of fish farmed in Israel.

F. GENERAL IMPRESSION

Fish farming in Israel is an outstanding example of the production of considerable amounts of valuable food rich in protein, where nature, left to itself, would hardly produce anything at all.

Since ponds have to be dug and water pumped into them, the water fertilized and the carp (constituting the bulk of the fish stock in Israel) fed, intensive farming is preferred to extensive farming. The water used for the ponds is slightly brackish, and therefore not suitable as drinking water or for irrigation. The ponds are very intensively used. They do not lie fallow periodically, and ploughing or harrowing their bottom is not done. Intensive farming entails, however, the risk of heavy attacks by parasites and diseases. Governmental stations provide the farmers with advice on how to control such pests.

The Israeli ponds in recent years produced the following quantities of marketable fish (in metric tons):

	1968	1969	1970
Carp	8 572	8 362	10 140
Tilapia	931	1 187	1 051
Mullet	597	680	675

The organizational form of the kibbutz makes it possible to use labour as efficiently as possible. It is no doubt still possible to increase the degree of mechanization in some aspects of farming, such as the spreading of food, fertilizer and pesticides. Mullet farming can only be expanded when fry can be produced under controlled conditions, as for carp and *Tilapia*.

G. TECHNICAL ASPECTS

G.1. Dipnets for catching mullet fry

A semi-circular iron frame, straight underneath, and measuring 50 cm in width, is attached to a wooden shaft. This frame is used to support a net made of nylon mosquito netting.

G.2. Hand-seine for catching mullet fry

The hand-seines are 3 to 5 m long and about 70 cm high. The headline is provided with floats made of cork or plastic, the ground-rope with sinkers. Sticks are attached to the lateral ends of the net. The net is operated by hand, the worker wading out in the shallow water of an estuary.

G.3. Composition of feed pellets

Pellets [a] for carps containing 25% crude protein are made up of:		Pellets [b] with 17% crude protein:	
Wheat	620 kg	Wheat	625 kg
Fish meal	150 kg	Fish meal	75 kg
Soybean cake	130 kg	Soybean cake	50 kg
Agglutinants	100 kg	Sorghum	250 kg
	1000 kg		1000 kg [c]

[a] Price US $ 100/metric ton.
[b] Price US $ 85/metric ton (additive included).
[c] To which is added 1 kg of the amino acid menthionine.

In the ponds with less than 700 kg/ha of fish, 100% sorghum can be fed to the carp; when there are 700 to 1 000 kg/ha, one should give 25% pellets and 75% sorghum, at 1 000 to 1 500 kg/ha 75% pellets and 25% sorghum, and whenever there are more than 1 500 kg/ha of fish, one should give 100% pellets, according to the feeding table.

G.4. Fish elevator

The fish elevator consists of a 15 to 20 m long sloping rail supported by a steel frame. A metal container with perforated side-walls slides up and down the rail, drawn by a gasoline engine. This container has a capacity of 70 l. Over the transportation tank its bottom is opened to dump the fish in the tank.

H. BIBLIOGRAPHY

Anonymous, 1972. The Grey Mullet. Induced Breeding and Larval Rearing. Oceanic Institute Waimanalo, Hawaii, Rep. No. 01-72-76-1, 202 pp.

Mires, D., 1970. Preliminary observations on the effects of salinity and temperature of water changes on Mugil capito fry. Bamidgeh, Bull. Fish Cult., Israel, 22 (1): 19—24.

Odum, W.E., 1970. Utilization of the direct grazing and plant detritus food chains by the striped mullet, *Mugil cephalus.* In: J.H. Steele (Editor), Marine Food Chains, Oliver & Boyd, Edinburgh, pp. 222—240.

Perlmutter, A., Bograd, L. and Pruginin, J., 1957. Use of the estuarine and sea fish of the family *Mugilidae* (grey mullets) for pond culture in Israel. Proc. Tech. Pap., Gen. Fish. Counc. Mediterr., No. 4, 289—304.

Sarig, S., 1970. Fisheries and fish culture in Israel in 1969. Bamidgeh, Bull. Fish Cult., Israel, 22 (3): 51—68.

Sarig, S., 1971. Fisheries and fish culture in Israel in 1970. Bamidgeh, Bull. Fish Cult., Israel, 23 (3): 63—84.

Sarig, S., 1971. Disease of Fishes. Book 3: The Prevention and Treatment of Diseases of Warm Water Fishes Under Sub-Tropical Conditions With Special Emphasis on Intensive Fish Farming. Tropical Fish Hobbyist Publications Inc. Ltd., Jersey City, N.J., 127 pp.

Yashouv, A., 1969. Mixed fish culture in ponds and the role of Tilapia in it. Bamidgeh, Bull. Fish Cult., Israel, 21 (3): 75—92.

Yashouv, A. and Ben-Shachar, A., 1967. Feeding experiments under laboratory conditions with *Mugil cephalus* L. and *Mugil capito* Cuvier. Bamidgeh, Bull. Fish Cult., Israel, 19 (2/3): 50—66.

Yashouv, A. and Berner-Samsonov, E., 1970. Contribution to the knowledge of eggs and early larval stages of mullets (*Mugilidae*) along the Israeli coast. Bamidgeh, Bull. Fish Cult., Israel, 22 (3): 72—89.

Chapter 2

"VALLICULTURA": MULLET FARMING ON THE ADRIATIC COAST OF NORTHERN ITALY

Example
Azienda Val Pisani, Ca' Venier, Rovigo (45010)
Director-Proprietor: Dr. Gino Ravagnan, Via Euganea 17, Padova (35100).
Period of observation: 1970—1972.

A. BACKGROUND

A.1. General principles

The littoral of the "Alto Adriatico", the Italian side of the northern Adriatic, is characterized by a series of lagoons and inlets, from Punta Sdobba in the north, to Punta Marina due northeast of the city of Ravenna. These lagoons are predominantly rather shallow, and contain brackish water of a salinity which may fluctuate between 10 and 35‰ in the course of the year. Several species of katadromous fishes, reproducing in the sea, enter these lagoons in spring to take advantage of these rich feeding grounds where high water temperatures prevail in summer. When water temperatures in the lagoon drop below those in the open sea, these fish (with the noticeable exception of the eel) tend to leave the lagoons, to return the next year, as soon as water temperatures become attractive again. When these fish, the eels included, approach sexual maturity, they invariably migrate to the sea to reproduce, irrespective of the actual water temperature. These seasonal migrations were known by the ancient Romans, and many centuries ago a system was developed to catch the scaled fish and the silver eels during their seaward migration. Close to the sea, a kind of weir, of poles and of grids of reeds, was constructed, to guide the fishes to a trap. The lagoons were not dammed completely, for communication with the sea, to let in the tides and the juvenile fish, born in the sea, was rightly considered of vital importance. These brackish bodies of water, provided with a fish weir in their seaward section, are called "valli", derived from the ancient word "vallum" which means "paling". In 1535 there were 61 of such "valli" in this region, and nowadays the total area of the "valli" is estimated to be 40 000 ha.

Eel and mullet are the fish of greatest commercial importance in the valli, but in addition some valuable carnivorous species may be found in limited quantities.

From olden days, the valli of the Alto Adriatico belonged to large landowners, who once used these waters for their pleasure, namely the chase, fish farming being considered, at that time, as an inferior occupation. The valli are still largely in the hands of large landowners, but their interest in the fishery aspects has now grown appreciably.

For ages the valli were worked with very simple, extensive methods: fry of eels and mullets would enter the warm water of the valli in spring, where they would grow and fatten on the abundantly available food. All that had to be done was to maintain the weir in good condition, making escape of larger fish virtually impossible, and to catch the fish in the traps in the periods of seaward migration. It was deplored that mullet would tend to leave the valli when water temperature dropped steeply in late autumn, long before they had grown to their optimal weight. If their voyage to the sea were made impossible, they would run the risk of being killed by low water temperature in mid-winter, unless they could find some deep well-sheltered sites in which to hibernate.

This system of extensive exploitation of the valli yielded from 80 to 150 kg of eels and mullet per ha per year, in which the eel production accounted for 50 to 60% of that weight, with a rather stable output year after year, whereas the mullet production fluctuated widely from 20 to 60 kg/ha. These variations can be explained in terms of ravages caused by severe winters and of noteworthy irregularities in the arrival of the fry.

People often tried to estimate the number of juvenile mullets coming in by watching them pass over a white board situated at the site where the sea water comes in with the flood. The fry of the different species of mullet were identified and it was observed that each of them has its own season of arrival. Concern about a lack of mullet fry migrating into the valli led to the practice of capturing the fry with fine-meshed nets; groups of two to five men wading out in nearby shallow coastal water, and transferring the fry carefully to the exploited valle. To reduce the risks of winter mortality, a certain section of the valle would sometimes be deepened to provide more shelter. This was about as far as the extensive system of "valli cultura" went, until 10 years ago. Though the method was both empirical and artisanal, this simple system of farming eels and mullets in brackish water was, and is, rightly considered as the most outstanding and important in Europe in terms of production, management and technique.

The "vallicultura" might have continued in the time-honoured way, returning a reasonable production for labour and maintenance costs, were it not that this industry met with serious reverses ten years ago. Three cold winters in succession, of which that of 1962—1963 was the worst, killed virtually all the mullet confined in the valli, and following that event, migration of fry was poorer than ever before. In about 1957 an ecto-parasitic copepode of the genus *Argulus* attacked the eels in the northern lagoons. It appeared to be a previously unknown species, to be described as *Argulus giordanii* by Brian, and it rapidly made its way south, to play havoc in the valli in the River Po delta in about 1965. In later years it also made its appearance in the famous Lagoon of Comacchio. Locally the losses among the eels were extremely high, and the catches of silver eels at the weir dropped to virtually nil.

These events were the main reason for making a serious effort to switch over from extensive to intensive farming. Various attempts were made to plant elvers and halfgrown eels, introduced from elsewhere, and to introduce mullet fry from other sections of the Italian coast. It was soon realized that planting of costly mullet fry in large bodies of water, in which the natural conditions, including predators and parasites, could not be controlled, nor could artificial food be supplied economically, would not lead to the desired consistent results. Therefore another system was tried: keeping the small fry in confinement in small basins, offering them pelleted feed, protecting them in special wintering basins against the effects of cold weather, and transferring them thereafter to somewhat larger basins, still giving pellet feed. Then planting them only much later in considerably larger basins, or in the valli themselves; the extensive shallow bodies of water from which they will once be harvested at the weir during their seaward migration.

Dr. Gino Ravagnan was one of the pioneers of this technical evolution. He altered his Val Pisani profoundly, digging a great number of basins and canals of various dimensions, and built a brood-house to rear small fish fry to a size large enough to be put in earth basins. The turning-point was in 1965. The great enthusiasm of the industrialist Dr. Ravagnan, backed by his solid finances, soon showed the way to higher and more consistent yields from the old valli, though with unavoidable initial investment, and noteworthy expenditure for feed and care. Dr. Ravagnan soon realized that more intensive farming unavoidably means greater risks of losses through parasites and diseases, and that only scientific assistance could pave the way to control of such pests. Since there was no governmental fisheries institute in the area to study the problems of an intensified valli-culture, Dr. Ravagnan decided to organize scientific research himself, and built a small laboratory on his premises. Dr. Ravagnan's enthusiasm lead to the establishment, on 27th April 1967, of the "Centro Ittiologico Valli Venete", to which 35 valli owners are now contributing, and of which Professor Dr. P. Ghittino of the Istituto Zooprofilattico Sperimental del Piemonte e della Liguria has become the scientific advisor. The biologist Dr. G. Arcarese directs the laboratory, and is assisted by the fisheries technician Mr. V. Brasola. A veterinarian specialized in fish diseases acts as consultant and visits the laboratory frequently.

Thus the valli owners make a joint effort to collect the basic information required for a rapid switch-over from extensive to intensive farming. The results of the scientific investigations on control of pests and diseases, feeding fish fry of various species, artificial reproduction of valuable and scarce fish species, and various technical aspects are brought to the knowledge of the participants in a special bulletin: "Bollettina informativo per i soci". Dr. Ravagnan acts as secretary of the C.I.V.V.

A.2. Biology of the species

Intensive farming focusses special attention on mullets, of which the following species occur in the valli:

Scientific name	Vernacular name	Period of natural migration
Mugil cephalus	Volpina, Mecchiatto	August—September
Mugil chelo	Bosega	October—November
Mugil auratus	Lotregano	October—November
Mugil capito	Bottolo	October—November
Mugil saliens	Verzellatta	June—July

The first three species are in great demand and bring the highest market prices. Unfortunately not all three species are numerous, *Mugil cephalus* even being quite rare. The mullets are predominantly herbivorous and are equipped with an extraordinarily long intestinal tract. Therefore, they are reckoned to be of the second link of the food chain. Their biology has been described in Chapter 1, dealing with mullet farming in Israel. An important feature of mullet behaviour, of which one can make an ample use in intensive farming, is their pronounced rheotaxis. They react promptly on currents and can thereby be attracted to the sites where they are to be concentrated for transfer or capture.

Two species of carnivorous fishes, respectively belonging to the Percidae and to the Sparidae, are also of practical importance: *Dicentrarchus labrax*, locally called "spigola" or "branzino" and *Sparus auratus*, locally called "orata". Just as the mullets, they tend to migrate into the valli, where they evidently find an abundance of food. They fetch a very high price on the fish market. Despite their carnivorous habits they can be trained to take pelleted feed, and some practical success has already been obtained in rearing their fry (gathered in nature) to marketable size, the latter species showing the more consistent results of the two.

The other fish species of great commercial importance for the valli is the European eel, *Anguilla anguilla*. The biology is well-known: its spawning grounds are to be found at considerable depth in the Sargasso Sea, and its *Lectocephalus* larvae migrate, carried by the Gulfstream, towards the European coast, a journey which lasts more than two years. On arrival at the continental shelf they metamorphose into transparent elvers, which try to make their way towards the estuaries. After having reached brackish or fresh water, they get the dark backside pigmentation so typical for the eel. Where ample food is available they grow fast, but only as long as water temperatures are 18°C or higher. Below 16°C growth is practically nil and in midwinter they hibernate in muddy bottom layers. It takes eels years, more in female than in male specimens, to reach the stage of sexual maturity. They then undergo a second metamorphosis, this time into "silver eels", the migratory form with its big eyes and silver-white belly, which begins its journey to the Sargasso Sea in late summer or autumn. They are then of prime quality, having accumulated a lot of fat to serve them as fuel for the long journey, and also for the production of eggs or sperm, for they do not feed at all after metamorphosis. When they leave the fresh and brackish water the gonads are developing, but eggs and sperm will only be produced later, unobserved by man.

During their sojourn in brackish or fresh water the eels lead a carnivorous life. Various invertebrates constitute their daily diet from the beginning, but some eels specialise later on fish as food.

A.3. Geographical situation

The Azienda Val Pisani is to be found in a bight of the Po di Maestra (Fig. 10), which branches off towards the north from the Po di Venezia in the River Po delta. The whole property, with a length of 5 800 m from west to east, measures 756 ha, of which about 500 ha consists of sheets of water, at least partly representing former courses of the River Po. The headquarters of the company are to be found at 45° 10′ N and 12° 22′ E, about 10 km east of the town of Contarina.

A.4. Hydrographic pattern

About 500 ha of the 756 ha property consists of water. Originally the whole valle was one big sheet of water. After being used for extensive fish farming for centuries, the first hydrographic intervention was made about 100 years ago, creating a deeper basin in which mullets and other vulnerable fish could retreat to hibernate. All the other modifications date from more recent years, starting in 1965. The original, much too large, body of water was thereby subdivided into three major sections, and in addition a great number of smaller basins of various dimensions were dug, some of which were especially designed for hibernation. They now cover together 100 ha. This process is still going on; the company operates with three draglines to carry out this arduous work. A network of channels provided with locks, slides and screens facilitates intercommunication. The three big sections, still used for extensive farming, have a depth of 60 to 100 cm. In reality, they represent old courses of the Po river, with bottom deposits of fine alluvial sand or clay, on which a layer of peat may have grown here and there. The level of the water is about 2.5 m below the average sea level, 15 to 20 cm higher in periods of heavy downpour. The whole area, characterized by a quaternary layer about 5 000 m thick, deposited on rather steeply sloping tertiary layers, is subsiding, a process which was greatly accelerated by ex-

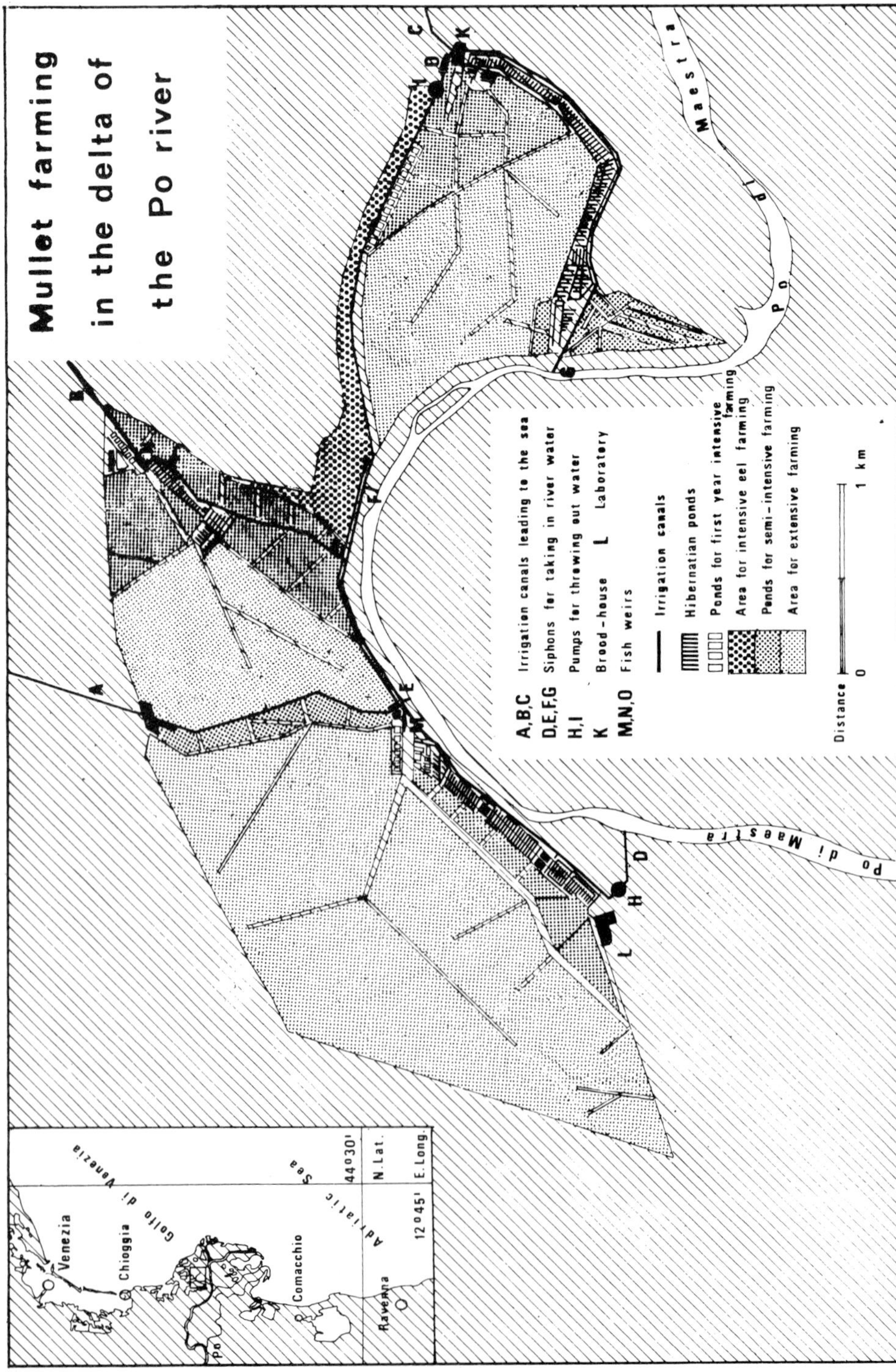
Mullet farming
in the delta of
the Po river
A,B,C Irrigation canals leading to the sea
D,E,F,G Siphons for taking in river water
H,I Pumps for throwing out water
K Brood-house L Laboratory
M,N,O Fish weirs
Irrigation canals
Hibernatian ponds
Ponds for first year intensive farming
Area for intensive eel farming
Ponds for semi-intensive farming
Area for extensive farming
Distance 0 1 km
Po di Maestra
Venezia
Chioggia
Golfo di Venezia
Adriatic Sea
Comacchio
Ravenna
Po
44°30' N. Lat.
12°45' E. Long.

Fig. 10. The Po di Maestra, which feeds the ponds with fresh water. (Photo P. Korringa.)

traction of a mixture of methane and water from deeper layers in the period 1950—1961. This led locally to a subsidence of 40 to 50 cm per year. Therefore the hydrographic situation is no longer comparable to that in the olden days, when the water level in the valli corresponded to the average sea level.

A sturdy dike enclosing the whole property prevents incursion of the sea, and since the Po di Maestra, bordering the area to the south, is also fringed by heavy dikes, the river cannot penetrate into the basins either.

To control both the water level and the salinity in all sections, three big irrigation canals have been constructed, leading to the sea, some 2 km away from the property. Inlet works, 5 m wide, provided with doors equipped with vertical slides, measuring 200 × 50 cm, make it possible to control the flow of sea water very accurately. The canals traverse the whole property, branching off here and there so that the sea water can be guided into all sections. Valves constructed in crucial places facilitate an appropriate distribution of the sea water. The fresh water needed is taken in from the Po di Maestra, which always has a higher level than the basins. Siphons (35 and 20 cm in diameter) on four sites on the northern bank of the river are used to obtain the water. Here too the quantities of water taken in can be regulated. The irrigation canals are used for sea water and fresh water; they often

Fig. 11. Digging new fish ponds in the Azienda Val Pisani. (Photo Dr. G. Ravagnan.)

contain a mixture of the desired salinity. Both sea water and fresh water can be let in by gravity, but pumps are needed to eliminate excess water from the low-lying basins. There are pumps on two sites of the property: at the western end of the northwestern irrigation canal, where three pumps with a capacity of 2 000, 800 and 700 l/s are installed in a special pump-house, and at the easternmost end of the property, where there is a pump with a capacity of 2 500 l/s.

The pumps are not in permanent use: the western ones are used for about 1 200 hours per year, especially in autumn and spring, when various basins have to be drained. In these periods the eastern pump renders assistance for 500 to 600 hours per year. In summer, the eastern pump is, above all, used for a system of internal circulation of the water, in which the large shallow sheets of water serve as areas in which the water leaving the rearing basins undergoes a natural microbiological purification.

Water temperature

The shallow bodies of water follow the air temperature faster than the water of the open Adriatic. In January, one can expect a minimum temperature of 2°C, and in severe winters there may even be a sheet of ice on the water. In spring, the temperature rises rapidly to reach 20°C in April, and to

Fig. 12. The northeastern section of the property with irrigation canal (B) leading to the sea, and the fish weirs (N) in the centre. At the right ponds for hibernation. (Photo Dr. G. Ravagnan.)

reach a maximum of 28 to 30° C in August. In September temperature drops under the 20° C level again.

Salinity

The salinity fluctuates considerably under the influence of precipitation and evaporation. Most of the time the salinity will vary between 28 and 30‰, lower than the adjacent sea water, but in periods of drought it may rise to 40‰, higher than that of the sea water, if river water is not siphoned

Fig. 13. The easternmost sector of the fish farm with irrigation canal (C) leading to the sea, with the fish weirs (O), and with centre-left, the new brood-house, open towards the southwest. (Photo Dr. G. Ravagnan.)

in in sufficient quantities. In winter, in periods of heavy downpour, the salinity may drop to about 15‰.

pH and oxygen

The pH of the water varies usually between 8 and 8.4, and the oxygen saturation usually gives no cause for concern, unless thick layers of the green weed *Enteromorpha* put the lower water layers in the dark.

Fertility

The water of the lagoon is evidently richer in plankton than the open sea, as can be concluded from its greener colour. The subsoil being a finely grained alluvium, brought down by the Po river, one can safely assume that it is quite fertile. Nevertheless, continued exploitation of the same area might lead to a gradual decline of the fertility of the uppermost bottom layers. Moreover, mud deposits may act as a kind of nutrient trap. Dr. Ravagnan considered that it might therefore be useful to plough the bottom of the large fish culture basins occasionally. He designed a special flat-bottomed boat (Fig. 14), equipped with a powerful compressor, which can blow air into a battery of 11 pipes, some 4 m long, articulating at the stern of the ship so that their ends can be let down on the bottom. This blow-boat is used to churn up the bottom layers. In spring, if possible, the whole area of the three large basins is treated this way. Where the bottom is soft, the air-tubes dig quite deep into it, and disperse any organic material deposited there. Anaerobic fermentation of organic material is thereby cut short, and rapid aerobic biodegradation ensures recirculation of the nutrients temporarily trapped in the bottom deposits. Dr. Ravagnan is convinced that his contraption contributes materially to the maintenance of the productivity of his fish farm.

Fig. 14. The blow-boat to churn up the bottom of the larger ponds, the pipes turned upwards. (Photo Dr. G. Ravagnan.)

Fig. 15. Details of the blow-boat with its compressor installation. (Photo Dr. G. Ravagnan.)

The weirs

Of crucial importance are the weirs, constructed at three well-chosen sites. The ancient type of weir, made from wood and reeds, has been abandoned, and instead batteries of locks have been constructed across the canal where water of high salinity meets the brackish water of the valli (Fig. 16). The locks are made of a series of longitudinal concrete walls provided with grooves at both ends, and also in their middle section. In these grooves fit metal slides across the locks, the slides being either solid or perforated. The bottom of the locks is also made of concrete, but care is taken to create a rounded-off slope where the wall meets the bottom, to ensure a better fit of the slides. In the lower part of the landward perforated slides there is a vertical slit provided with adjustable rubber flaps. Fish migrating to the sea can pass through this slit, pushing the flaps aside, but will not be able to return (Fig. 17). In the season of eel migration the slit is made narrower than at the time the mullets migrate seawards. This entrance can be closed, if so desired, with a narrow metal sheet, without removing the whole slide.

This contraption makes it possible to arrange free access of the sea water but to prevent free migration of the fishes. Rheotaxis of the fish or their desire to reach water of a certain temperature or salinity easily leads to their concentration within the locks, from which they can easily be seined out quantitatively, even the slippery eels.

Fig. 16. The battery of modern fish locks makes escape of fish to the sea impossible. (Photo P. Korringa.)

Fig. 17. Fish-lock slide with adjustable slit with rubber flaps. The slit can be closed with a metal sheet. (Photo P. Korringa.)

A.5. *Legal aspects*

The Azienda Val Pisani belongs to Dr. Gino Ravagnan, but is only part of his private property. He owns in addition a vast area of arable land in the same region, and has interests in several industrial enterprises.

The personnel of the azienda are in his service, but the costs for scientific research are shared by all the associates of the Centro Ittiologico Valli Venete (C.I.V.V.). A special aspect is the present political tendency in Italy towards expropriation of large landownership. A modern method of development, leading to a high productivity and offering employment to many people, may reduce such a threat.

B. SEQUENCE OF OPERATIONS IN INTENSIVE FARMING OF MULLETS

B.1. *Procuring the fry*

Long ago the natural immigration of mullet fry, though variable, sustained the mullet population of the valli. When, based on observations at the weir, it was thought that the numbers of juvenile fish making their way to the valli were insufficient, additional numbers were fished in the adjacent coastal waters, and transferred carefully to the brackish water behind the weir. This procedure no longer leads to the desired result, for the catches made with fine-meshed nets in the shallow coastal water of the region are nowadays extremely poor. It is therefore no longer possible to stock the valli with fry caught in the neighbourhood.

The reason for this serious reduction in the numbers of fry arriving in the coastal waters of the Alto Adriatico is not fully clear. Some think pollution is to be blamed, but cannot indicate which kind of industrial pollution would be responsible for the disappointing catches in the entire coastal area. Moreover, the mullets spawn well offshore in unpolluted water, and if their small fry are brought in from elsewhere, they survive and thrive well in basins filled with coastal sea water, which certainly would not be the case if this water were seriously polluted. Hardly any research has been carried out to investigate the problem. There are reasons to assume that the parent stock, ready to spawn, might be much smaller than before, because the modern weir constructions give fewer, if any, mature specimens a chance to regain the sea where they were born. If it could be demonstrated that this is the main reason for the scarceness of natural fry in the Alto Adriatico, measures could easily be devised to counter it.

For the time being, the paucity of fry in the area is considered an irremediable fact, and since artificial reproduction of mullet has not yet been achieved on a technical level, the only way to acquire mullet fry is to buy it from elsewhere. Since the mullet farmers are inclined to pay a good price for mullet fry, fishermen in various places on the Italian coast have specialized in catching it, either on foot, in shallow coastal water and in estuaries, or from a boat, using fine-meshed seins or square nets. The dealers have tankers equipped with oxygen installations and bring the fry to the farmers over considerable distances. The farmers of the Alto Adriatico receive their mullet fry from Taranto, from Toscana and from various other centres on the Italian coast, both Adriatic and Mediterranean. They pay a high price for mullet fry: from L.It. 10 to 20 (freight included) per specimen of *Mugil chelo* or *Mugil cephalus*, and at that price the company under consideration is willing to buy any quantity offered. Upon arrival the juvenile mullet have to be checked to ensure they do indeed belong to the species one desires to acquire. It is considered as normal that a certain percentage will die (25 to 35%), predominantly in the first few days and weeks after arrival, as result of the hardships of being caught, transported and transferred brusquely to new environmental conditions. If dealers offer fry of the carnivorous species *Dicentrarchus labrax* and *Sparus auratus*, these are purchased as well, to be reared separately; the former fetch L.It. 20 per specimen, the latter as much as L.It. 150 to 200.

B.2. Rearing the fry

Mugil cephalus ("volpina" or "mecchiatto") reproduces in offshore waters in August, and its fry, measuring 15 to 20 mm in length, can be purchased in October and November. Sometimes fry of this species can be bought in February and March, when they measure about 25 mm.

If rather expensive fry are bought, it is for economic reasons imperative to rear that under controlled conditions in small basins, and to refrain from putting them straight into the big valle with its many dangers. The Azienda Val Pisani has a covered brood-house (Fig. 18) at its disposal, in which are two rows of 20 circular concrete basins, 4 m in diameter. These basins are shallow, being deepest in the centre, and they receive a constant supply of filtered sea water, let in by gravity from a large outdoor basin. This water can be let in as a simple jet, but for rearing small fry, it is guided through a perforated pipe, placed radially, which squirts it, in about 100 thin spouts, obliquely on the surface of the water, at a total rate of 1.5 l/s. This water serves to exchange the water in the basin and to keep it in a slow circular motion. This keeps the young fish swimming, but if they wish they can rest on the bottom in the central part of the basin. Debris is automatically collected in the centre and can be washed away through a grating leading to an outlet, simply by opening a valve.

In such a 4 m circular basin up to 150 000 specimens of mullet fry of 10 to 20 mm length can be reared. The big advantages of such a system are easy observation, constant flushing, automatic feeding, and application of disinfectants, both on a regular basis and incidentally when diseases or parasites

Fig. 18. Interior of the new brood-house with its circular basins and oblique spouts of water. Right: the "Capo Valle". (Photo P. Korringa.)

threaten to spread. In principle each basin with fry is disinfected every 10 days, interrupting the flushing for a short while. In general, the fish disinfectant Neguvon-Masoten (Bayer) is used. In special cases one makes use of malachite green. For automatic feeding, a feed-distributor with a slowly moving conveyor belt is used (Fig. 19). It runs for 2.5 h by clockwork. The dry feed is hereby spread very gradually into the basin. As food, pellets made by the company SOIMA (Societa Italiana Mangini) at Cologna-Veneta are used, and the young fry are supplied for the first 20 to 30 days with the finest brand called "starter". Thereafter, when they have reached a length of about 30 mm, a coarser brand is given, "No. 1". The fry are fed during daylight only, and not at all on Sundays. The quantity is estimated empirically: just that much as is eaten eargerly. Juvenile fish should receive daily, 5 to 7% of their weight in pellets, the exact quantity being dependent on water temperature. The percentage can be lowered as the fishes grow.

Young fishes can be reared in the brood-house until they have attained a length of about 8 cm, but then of course, in smaller numbers per basin. As the fish grow they should be distributed over a greater number of basins, and

Fig. 19. Automatic feed distributor with slowly moving conveyor belt. (Photo P. Korringa.)

at a length of 8 cm, there should be no more than 13 000 specimens in one basin.

For various reasons they are not usually kept that long indoors, but are transferred to outdoor basins from a length of 5 cm upwards. Since the various species show different growth rates and since fry are purchased in various parts of the year, the basins can be used efficiently.

The brood-house ("avannotteria") shows an open facade facing south, which can be closed in winter with transparent panels. In winter it can be heated by a plant producing 600 kcal/h. Then one can switch over to recirculation of the water, making use of a modern filter installation. Warm water discharged from the brood-house is always led into winter ponds in an effort to keep their temperature on an adequate level.

B.3. Transferring the fry to outdoor basins

For the second rearing phase rectangular earth ponds, measuring about 600 m^2, are used, with a depth of 1.5 to 2 m. If such basins are well-flushed, about 2 kg of fish can be kept per m^3 of water. At a water temperature of 15°C a well-stocked pond should be flushed in 24 h, at the summer maximum of 25 to 28°C, in 6 h. This requires an influx of 15 and 60 l/s, respectively.

To reduce the change of habitat to an absolute minimum, the earth ponds are usually equipped with a "fore-pond" made of concrete, measuring some 20 m^2, communicating with the earth pond, but separated from it by a screen. Here they receive their pellet feed. After some time, allowing final observations on their actual number, their good health and active feeding, the screen is pulled up so that the juvenile fishes can spread over the earth pond. They will frequently come back to the "fore-pond" to find their pellet feed, for there is installed the food distributor. They receive their daily dose of pellet feed of a grain size corresponding with their own size, and in quantities calculated with the aid of feeding tables, taking their number and weight, and also the actual water temperature into consideration.

B.4. Hibernation ponds ("peschiera")

Mullet cannot stand low water temperatures. That is why they want to leave the valli in mid-winter, to regain the temporarily, warmer sea, and why heavy losses may occur in cold winters when the weirs make their escape impossible, and the valle does not offer deep sites to hibernate rather safely. Arranged in order of major resistance to low water temperatures the species found in the valli are: *Dicentrarchus labrax*, *Mugil saliens*, *Mugil capito*, *Mugil cephalus*, *Mugil chelo*, *Mugil auratus*, *Sparus auratus*.

Practical experience shows that the valli in the Alto Adriatico cannot possibly yield a regular crop of mullet when they do not offer the valuable species a "peschiera" — a safe refuge in winter. Purchase of costly fry and introduction of a system of intensive culture make safe hibernation, for economical reasons, a prerequisite. In the Val Pisani, the azienda of Dr. Ravagnan, a great number of hibernation basins have been dug, principally in recent years. The most important features of such hibernation basins are a considerable depth and adequate shelter against cold winds. These basins have a length of 30 to 100 m, a breadth of 3 to 6 m and a depth of 4 to 5 m. They are constructed in a direction perpendicular to that of the cold Bora wind, which blows from the northeast. The banks of the basin should be about 1 m above the water level, and whenever possible, evergreen shrubs are planted on the banks to offer extra shelter. Percolating phreatic water ensures stable water temperatures in the deeper parts of the basin, and in periods with snow or frost, the water of the wintering basin can be protected still further by creating a floating toplayer. This thin toplayer (30 to 50 cm) of low salinity water will cool off rapidly, but is flushed out before it can affect the temperature of the deeper layers of the basin.

Differences in water level, and a good system of intercommunication, with screens, slides and valves to bring water in the desired quantities to the sites where it is deemed necessary, are characteristic for the equipment of the hibernation installations. In special cases powerful pumps can be employed.

In November, the fish of vulnerable species are transferred to the wintering basins. They stay there until the second part of March and will not receive pellet feed in that period. Whenever possible, fish should not be taken from rearing ponds with nets for transfer to the wintering basins. Since mullet react promptly to currents and temperature differences, it is usually easy to lure them to the wintering basins. The inlets and outlets of the outdoor basins are designed in such a way that the direction of the current can be reversed. The same principle can be used in spring, when one wishes to transfer the fish from the wintering basins to the larger basins, for semi-intensive farming; a jet of warmer water via the communication channel will promptly attract them. Adequate construction and appropriate manipulation of the hibernation basins does not completely eliminate the risks in really severe winters. Covered basins and artificial heating would mean a real improvement, but at too high a cost to defray the occasional losses.

Since the fish do not feed at the low temperatures prevailing in the wintering basins (10° C), and require much less oxygen than during summer activity, one can place up to 2 kg of fish per m^3 in them. In some newly constructed

Fig. 20. Mobile tank for transport of fish over short distances. Front side with pump and hose to fill the tank. (Photo P. Korringa.)

concrete wintering basins, with recirculated water of 7° C, it is thought possible to store up to 50 kg of fish per m^2 safely.

B.5. Basins for semi-intensive farming

In early spring the fish are guided to the basins for semi-intensive farming. The total area used for these basins must be quite large in comparison with that used for intensive farming in the first year. Practical dimensions are 40 × 20 m, the ponds made in series. These basins are quite shallow, 50 to 70 cm over most of their surface, but along the entire circumference they are provided with a channel 1.5 to 2 m deep. The shallow part is used as pasture, the deeper channel as a refuge in bad weather. The basins are continuously flushed, in the warmer part of the season even quite intensely: 50 to 60 l/ha/s, which makes it possible to stock the basins in a density up to 2 kg of fish/m^2.

Pumps can be used if the ponds must be emptied. The fish congregate then in the deeper ring-channel, and finally in a small, still deeper section from which they can be fished out with a net. The term "semi-intensive farming" is used about this stage because the fish depend here partly on natural food and partly on pellets. The dose of pellets administered (e.g. of the brand Camilotti, Padova) is much smaller (0.5 to 1% dry pellets, calculated on net-weight of the fishes) than in the basin for intensive farming, but the herbivorous mullets find ample natural food (e.g. certain algae, detritus) in the shallow central part of the basin. No manure or inorganic fertilizer is used to further the production of natural food in these basins. The pellets are scattered in the shallow central part of the basin by a tractor equipped with a blower (Fig. 21).

B.6. Harvesting the fish from the ponds for semi-intensive farming

At the end of their second season of cultivation, some of the mullet will already have reached a weight of 300 to 350 g, which means that they are marketable. Therefore the population of the ponds for semi-intensive farming is not guided quantitatively towards hibernation ponds when cold weather sets in, but is taken up with a net from the deeper fishing sites after draining the pond. The fishes are then quickly sorted and those having attained the marketable size are kept separate, the others being transferred to a "peschiera", a hibernation pond.

The marketable fish are either marketed directly, or temporarily stored in a small, deep basin, to await a period with top prices. As a rule the percentage of mullets achieving consumption size after some 18 months of farming is rather small. The majority of the fish are always transferred to a wintering basin.

Fig. 21. Blowing pellet feed into the ponds for semi-intensive farming. (Photo P. Korringa.)

B.7. Extensive farming

From the wintering basins the mullets are transferred to shallow basins for extensive farming; very large sheets of water, including the three big sections which are "ploughed" each spring with the blow-boat. Here the fish depend entirely on natural food. No pellets or fertilizers are administered. The quantities of fish grown per hectare are reduced correspondingly. For every hectare of intensive cultivation there should be 100 ha available for extensive farming. Continuation of semi-intensive farming after the second hibernation could be considered, but the quantity of pellets to be given increases with the size of the fish. Moreover, one has to take into account that larger fish are less vulnerable to the attacks of predators, and in many cases, also to the ravages caused by parasites, than are juvenile specimens. This reduces the need to rear them under controlled conditions. The losses in the ponds for semi-intensive farming are estimated to be in the order of 5% per year; there is no reason to assume that they will be much higher during the sojourn of the larger fishes in the extensive farming ponds.

Fig. 22. Typical boat of simple construction as used by fishermen operating in the valle. (Photo P. Korringa.)

B.8. Harvesting the marketable fish

It would be no easy matter to harvest the mullet from the valli (the large ponds for extensive farming) if the water had to be pumped out and if the very agile fish had to be collected with nets from some deeper sections. In reality nature takes care of harvesting. When approaching sexual maturity, which takes a different number of years in the various species, the mullet migrate seawards. They then try to make their way through the weir. The modern construction of the weirs (locally called "lavorieri") of the Val Pisani fish farm is such that the mullet will pass the slits in the landward slides equipped with the rubber flaps, but will have no chance at all to escape to the sea. It is an easy matter to fish them quantitatively from the locks in the weir with a small seine net (Fig. 23). They can either be marketed directly or be put temporarily in a special storage basin ("risalvi"), awaiting a better market.

B.9. Marketing the mullet

The fish to be marketed are washed but not stripped, and packed in plastic boxes. Mullets weighing 600 to 800 g each, are over 3 years old, those of a length of over 40 cm, are 5 to 6 years. The fish is always sold as fresh fish and the dealers come to collect them on the spot. The price fluctuates with

Fig. 23. Fishing with a fine-meshed seine net in a fishlock. (Photo P. Korringa.)

the time of the year, with the quantities offered, and also with the quantity of sea fish landed in nearby Chioggia. Therefore, it pays to store the mullets alive in small ponds if the actual price is considered to be too low.

B.10. Other species of scale fish from the valli

Of the scaled fish, mullet form by far the largest share of the annual crop; mullet of different species with a different price per kg and different seasons of seaward migration. Of the carnivorous fish, the orata (*Sparus auratus*) and the spigola (*Dicentrarchus labrax*) are held in high esteem. The orata may reach a weight of 150 to 180 g in one year of cultivation, and attain the marketable size of 400 to 500 g in the second year of farming. It feeds predominantly on cockles (*Cardium edule*) and does not interfere with mullet living in the same body of water. It brings a very high price (L.It. 2 500—2 800/kg). Therefore its fry, measuring 15 mm, is much in demand, but unfortunately it is very scarce.

The spigola has about the same market price as the orata, but it takes this species more years to attain the marketable size and it suffers a higher mortality rate. The spigola is a very voracious fish, which could destroy great

Fig. 24. Fine-meshed seine net used to collect fish concentrated in the fish locks. (Photo P. Korringa.)

numbers of smaller mullet. It is therefore preferably grown in intensive farming on moist formula food or on ground fish, and when grown in semi-culture, one takes care that the mullet in the basins under consideration are large enough to avoid interference.

A little fish, abundant in the valli, and reproducing there, is the "latterino" (*Atherina boyeri*), which attains a size of 8 to 10 cm. It is no doubt an important prey for the carnivorous species, but is also collected on some scale for human consumption. It is a by-product of minor value.

B.11. Eels

Of great commercial importance for the fish farms of the Alto Adriatico is the eel, *Anguilla anguilla.* In principle, the eel production is still based on the old extensive method of cultivation. Formerly the eel production in the Val Pisani amounted to an average of 50 metric tons per year, but the disastrous mortality caused by the *Argulus* outbreak in the early nineteen-sixties brought the migration of silver eels down to an unprecedented minimum of only 30 kg in 1965. Gradually the production increased again, and now each year 10 to 12 metric tons of eels are harvested from the Val Pisani. The

Fig. 25. Mr Brasola demonstrates a type of dip-net used to scoop fish out of the seine net in the fish lock. (Photo P. Korringa.)

natural immigration of elvers is no longer sufficient to supplement the stock in the valle. It is claimed that virtually no elvers are making their way into this valle now, though the Lagoon of Comacchio, somewhat further south, still receives a fair supply along the natural pathway. It is not clear whether the decline in the numbers of elvers is due to non-arrival on this section of coast or entirely, or at least partly, due to lack of facilities for their immigration. The old weirs of poles and reeds formed no barrier at all for the migrating elvers, but the concrete locks, with tightly fitting metal slides in the canals communicating with the sea, form an almost hermetic barrier for the immigrating elvers, and even where some would conquer this difficulty, they would meet further tight barriers on their way to the valle. Though it would not be difficult to build special devices to facilitate passage of elvers, this has not yet been done, presumably because it is assumed that the number of elvers reaching the coast is too small to merit any action.

As the natural immigration is nowadays absolutely insufficient to produce a fair quantity of marketable eels, the stocks are obtained from elsewhere. Elvers can be purchased from specialized dealers. In Toscana there are installations in the rivers of trap elvers. This is not primarily to catch elvers for stocking other areas, but to catch elvers to be eaten. In Spain and Italy the public is very keen on a dish of elvers, and is prepared to pay a high price for it. The dealers sell the elvers L.It. 5 000 to 10 000 per kg, and the consumer has to pay still more. If elvers are bought from the dealers in Pisa or Livorno, this price therefore has to be paid. If it were certain that a fair percentage would survive till metamorphosis to silver eels this price would be reasonable, for every kg of elvers contains some 2 500 to 3 000 specimens. The Val Pisani buys now, every year, 30 to 40 kg of elvers, but this not for direct planting in the valle, but for small-scale experimentation with intensive culture based on the Japanese experience in eel farming. The young eels are kept in small ponds and fed with the type of paste, sticking together, developed in Japan. Since the eels hardly grow at all at water temperatures of 17°C and lower, which virtually limits the growing season to the period May—October, a heated rearing house is now being built for rearing elvers in intensive culture, either to commercial size or at least to a size for planting in the valle. All this is still in the experimental phase.

However, the Val Pisani is stocked with eels of somewhat larger size, measuring 20 to 25 cm, of a weight of 10 to 25 g each. These are purchased either in southern France (Marseille), and transported by tanker, or from the 3 000 ha Laguna di Orbetello, on Italy's west coast, at a latitude of 42° 27′ N. The planting of half-grown eels in the Val Pisani has been carried out on an appreciable scale, and it seems certain that of the tonnage caught now (10 to 12 metric tons of silver eels per year), most, if not all, originate from planted stock.

C. FARMING RISKS

C.1. Hydrographic conditions

Since mullet cannot survive in water of very low temperature, severe winter may play havoc with the stocks, especially when no adequate hibernation ponds have been constructed. The losses in the winter 1962—1963 were so great that only 4 metric tons of mullet were caught in 1965.

C.2. Predators

It is difficult to establish the degree of predation upon the young mullet migrating naturally into the valle, and to ascribe given percentages of losses to identified predators. Based on observations on the number of mullet migrating into the valle in the period before 1965, and on the number of adult fish caught later, while migrating seawards, one could safely assume that less than 10% of the immigrants reached the adult stage. The species *Mugil chelo* and *Mugil cephalus* showed perhaps a somewhat higher figure. Among the predators on the juvenile stages are birds such as herons, little egrets, terns and gulls, but also the grass-snake and carnivorous fish, among which, in all probability are the larger eels.

C.3. Parasites and diseases

The explosion of *Argulus giordanii* in the early nineteen-sixties taught how serious losses caused by ectoparasites can be, even under conditions of extensive eel farming. As happens so often after such a devastating outbreak, one leading to serious reduction of the fish stock, conditions became gradually better, so that the production has again reached a level of about 25% of that before 1960. It is not clear in how far this should be ascribed to a general reduction of the eel stock, or to loss of virulence of the parasite, which was said to be a newcomer in this area when it suddenly increased, almost astronomically, in numbers.

In intensive farming one can take measures of control, but in a large basin for extensive farming, chemical control is virtually impossible. The mullet fry is disinfected upon arrival and during intensive cultivation the treatment is repeated at fixed intervals. Elvers and eels introduced from elsewhere are invariably treated with special disinfectants before being planted in the valle.

C.4. Theft

The large fish do not suffer appreciably from predation, but a more serious problem is theft and poaching. The property is very large, and especially with the three huge sections for extensive farming, is difficult to supervise. It happens repeatedly that poachers come with boats and nets in the night, and the fact that one of the three large sections invariably produces less per hectare than the two others should surely be ascribed to poaching. Occasionally even fish are stolen from the ponds used for semi-intensive farming. To keep these practices down, the company has three armed surveyors (guardia giurata) among the personnel, who take the nocturnal supervision in turn. Poaching for game and fish is inherent to large landownership in many a country. Simple country folk have difficulty in accepting ownership of edible wild creatures by a remote landowner from a higher level of society.

D. ECONOMIC ASPECTS

D.1. Rents

Since the director of the azienda Val Pisani is the owner of the ground he need not pay any rent, but has, of course, to pay taxes for his property.

D.2. Inventory items

The rapid switch over from extensive to intensive farming makes it necessary to use heavy equipment for digging ponds and canals. Now three draglines and one bulldozer are operated, but when all the basins are completed according to the plans, it may be sufficient to use one dragline only to keep the canals and basins at the desired depth. The heavy equipment is the property of the company. For permanent use are the two houses where the fishermen can rest, eat and shelter, the adjacent store-house for material, the garage and workshop. The new brood-house with its circular concrete basins will have a long life, and so will the eel-rearing house equipped with heating system, now under construction. The company operates with one 2-ton-truck, one van for transport of fish and with a truck equipped with a blower for the mechanized scattering of feed pellets for the mullets in the larger basins. For transport of live fish on the property, a specially constructed tanker is used. Further inventory items are the pumps of various capacity, partly electrical, partly diesel. Apart from the four larger pumps used to throw out excess water, there are several pumps of more modest capacity (e.g. one of 200 l/s) for local adjustment of water levels, and in summer also for ensuring an adequate oxygen content of the water in various basins. Further, there are the "lavorieri", the modern weirs with their slides equipped with fish migration slits. The fleet consists of the blow-boat equipped with a compressor, used for ploughing the large basins for extensive farming, and six small wooden row-boats. A certain number of seine nets for fishing out basins and weirs should also be placed on the list of inventory items. The laboratory with all its equipment, including pumps, filters, and aquaria of various dimensions, should be listed separately since it is operated jointly with other fish farmers by the "Centro Ittiologico Valli Venete".

D.3. Expendable items

The items predominating in the list of expendables are: feed, fuel and electricity. Feed pellets of various grain size are purchased for the fishes of different age. On the property there are three special pellet silo's for storage of the feed. They belong to the factories furnishing the pellets. The total quantity of feed used per year fluctuates between 200 and 300 metric tons. The price of pellets for juvenile fish is about L.It. 250 per kg, for the larger fish the price of the feed is about L.It. 100 per kg. The special formula food for eel and spigola costs, however, about L.It. 250 per kg. The prices for feed are rather unstable, greatly influenced by the price of fish meal on the world market. The larger the fish, the more feed they require per individual. Therefore, intensive farming in the brood-house, during the first few months after arrival of the fry, costs relatively little in feed, expressed in total tonnage. In the larger basins for semi-intensive farming, only part of the food is given in the form of pellets, the remainder being produced by nature. In some cases, as in farming the spigola, one feeds with ground fresh fish. Fuel is required for the draglines, the bulldozer, the blow-boat, for the various vehicles and for the smaller pumps. Electricity is required for operating the larger pumps. It is the "Capo Valle" who decides when to set the pumps in operation. Items of minor importance include the chemical products used for controlling pests.

D.4. Purchase of fry and eels

The production of mullet depends now entirely on an initial phase of intensive farming, followed by a semi-intensive phase, and after that partly on extensive farming. Natural immigration does not play any part now in the Azienda Val Pisani. Mullet fry costs now L.It. 10—20 per specimen, at least for the best species *Mugil chelo* and *Mugil cephalus*. The number of juvenile mullet purchased in the year 1969—1971 fluctuated around 1 000 000 specimens. Much smaller numbers of orata (*Sparus auratus*) were purchased: 50 000 to 100 000 specimens, and of spigola (*Dicentrarchus labrax*): 80 000. Orata fry cost L.It. 150—200 each, fry of spigola L.It. 20. The former grows rapidly and promises more consistent results. For experimentation with intensive farming of eels, in recent years 30 to 40 kg of elvers were purchased for a price of L.It. 8 000 to 10 000 per kg, and for stocking the large basins for extensive culture, every year 500 to 1 000 kg of young eels of 10 to 30 g each are purchased at L.It. 800—1 000 per kg. Several years must pass before the crop of eels and silver-eels trying to make their way to the ocean can be harvested.

D.5. Sales

After the change to intensive farming of mullet, followed by semi-intensive culture in larger basins, the production of marketable fish rose rapidly. In 1970 about 22 metric tons of mullet were sold, in 1971 over 82 metric tons and in 1972 about 90 metric tons. The price of the mullet varies according to the species and the seasons: for the better species *Mugil chelo*, *Mugil cephalus* and *Mugil auratus*, from L.It. 1 000 to 1 400 per kg, for *Mugil capito* L.It. 800 to 1 000 per kg, whereas for *Mugil saliens* a price of about L.It. 600 per kg is made. *Mugil auratus* (lotregano) cannot stand low water temperatures at all. Therefore specimens of 300 to 400 g, a weight reached after 3 years farming, are quite rare, but much in demand. One always sells them before the end of the year. The carnivorous species *Sparus auratus* and *Dicentrarchus labrax* yield from L.It. 2 500 to 3 000 per kg. The small latterino's (*Atherina boyeri*) is sold for L.It. 400 to 600 per kg and the total production varies from 1.5 to 2 metric tons per year. Of considerable importance is the sale of silver-eels, for which, the price fluctuates around

L.It. 1 000 per kg. The production is now rising rapidly, due to planting of young eels in the larger basins used for extensive farming. The eel production, which came down to a very low level some 6 years ago rose again to 10 or 12 metric tons around 1970, and is expected to reach 20 metric tons soon.

D.6. Market

The market prices fluctuate not only simply according to the seasons, but also under influence of the quantities offered for sale, and of the amount of fish landed from the sea. Though it is possible to keep mullet alive in small storage basins, awaiting better prices, it is desirable to study the marketing possibilities in greater detail. If more valli culturists switch over to intensive farming, and if sufficient fry can be purchased, one should expect a greater tonnage of mullets on the Italian market, perhaps even a tenfold increase on the present production. It must be possible to market them in a fish loving country like Italy, but one of the requirements will be an even distribution over the seasons, irrespective of migration peaks. Therefore deep-freezing of individual fishes as is done with salmon and rainbow trout in Norwegian fish farms would be a necessity. A fool-proof method of deep-freezing, guaranteeing no loss of quality during storage, has to be developed.

D.7. Personnel

On the azienda Val Pisani there are now 36 workers in employment. This high figure should be explained in terms of the switch-over from extensive to intensive farming. Most of the workers are engaged in creating new basins and canals. For the pure fish farming operations, only 15 to 20 hands would be needed. This is still quite a high figure in comparison with companies engaged in extensive farming only. The latter employ about eight hands for 500 ha. The combination intensive-extensive has evidently led to doubling of the personnel. In times of heavy work-load, the director occassionally calls in workers from his agricultural enterprises in the area. The wages vary from L.It. 100 000 to 150 000 per month, somewhat higher than those paid in agriculture in this area. The work on fish farms is often harder and more tiring.

D.8. Economics of intensive/extensive farming

It is clear that old style extensive farming is not attractive any more, since both the elver run and the immigration of mullet leave so much to be desired. A modern fish farm should be able to aim at a constant production of mullets and eels. The greater the importance of the intensive farming phase, the better one can control the production of marketable fish.

Intensive farming means not only that a brood-house and a great number of ponds and basins of various sizes, including wintering basins, are needed, but also that considerable expenses for feed and care have to be borne. The small fish need little food but considerable care. When the quantity of fry to be purchased is limited, and if such fry are expensive, it will always pay to rear them for at least some time in intensive culture under controlled conditions, leading to a much higher survival rate (Fig. 26). Mullet can be grown in intensive and semi-intensive farming to marketable size, but in the company under consideration, the mullet which are not yet marketable are transferred after their second season of growth to the larger basins for extensive farming. If losses remain at a low level, about 5% per year in these larger sheets of water, and all maturing specimens, in due course, turn up at the weir, it may, indeed, be economically sound to follow such a procedure. Quite another approach is, however, that though larger fishes may require more feed per individual than smaller ones, the conversion rate of pellet feed into fish flesh is usually better in the larger fish than in the juvenile specimens. It is a question of economic calculation whether it is wiser to continue semi-intensive

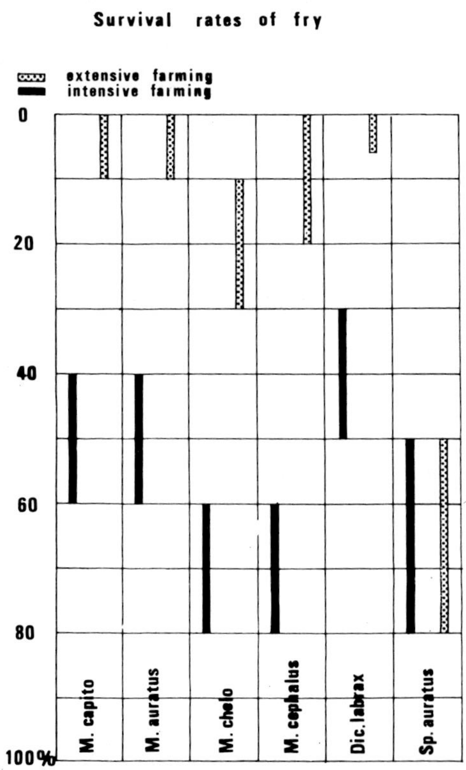

Fig. 26. Survival rate of fry in intensive and in extensive farming in the Azienda Val Pisani. (From Dr. G. Ravagnan.)

TABLE I

Purchase of fry, production of yearlings, production of older fish, Val Pisani 1969—1972

	Year	Number of fry	Rearing technique	Tonnage of yearlings (kg)	Individual weight of yearlings (g)	Survival of yearlings (%)	Tonnage older fish (kg)	Individual weights of older fish (g)	Survival of older fish (%)	Further growth (kg)	Market (kg)
Mullets	1969	1 200 000	Earth ponds	18 000	*M. cephalus* + *M. auratus* 70—100	40—50					
	1970	850 000	Earth ponds	12 000	*M. chelo* *M. capito* *M. saliens* 15—20	80	72 000	*M. chelo* 150—300 *M. cephalus* 280—300 *M. auratus* 130—140	95	50 000	22 000
	1971	600 000	Broodhouse followed by earth ponds	12 600		over 70	130 000			48 000	82 000
	1972	520 000	Broodhouse followed by earth ponds	12 500		80					
Orata	1969	100 000	Large semi-intensive ponds	10 000	120	80					

	1970	74 000	Large semi-intensive ponds	5 280	120	60	15 000	350—400	95		15 000
	1971	45 000	Broodhouse followed by earth ponds	4 350	120	80	15 000	350—400	95		15 000
	1972	7 000	Broodhouse followed by earth ponds	600	120	75					
Spigola	1969	70 000	Large semi-intensive ponds	280	20	20					
	1970	67 000	Large semi-intensive ponds	300			5 000	varies widely		5 000	
	1971	100 000	Broodhouse followed by earth ponds	6 200			7 200			7 200	
	1972	150 000	Broodhouse followed by earth ponds	1 200		40					

farming for one more year, or to switch over to extensive farming when the fish have reached a certain size.

The great success of intensive farming under conditions of scarcity of fry can be demonstrated from the figures of the azienda Val Pisani bearing on recent years (Table I). The table clearly shows that though the number of fry purchased showed a noteworthy decline from 1969 to 1972, the production of marketable fish ensuing, increased as the production methods became more intensive.

With the present technique of intensive farming applied in the azienda Val Pisani, the survival percentage in the first year of rearing ranges from 40 to 60% in *Mugil capito* and *Mugil auratus*; from 60 to 80% in *Mugil chelo* and *Mugil cephalus*; from 30 to 50% in *Dicentrarchus labrax*; and from 50 to 80% in *Sparus auratus.* If the fish are introduced directly into basins for semi-intensive or extensive farming, the survival percentage is estimated at 10% for *Mugil capito* and *Mugil auratus*; 10 to 30% for *Mugil chelo*; 1 to 20% for *Mugil cephalus*; 1 to 6% for *Dicentrarchus labrax*; and 50 to 80% for *Sparus auratus.*

E. GOVERNMENT ASSISTANCE

The Italian Government did not play an active part in the switch-over from extensive to intensive farming in the valli-culture of the Alto Adriatico, neither by subsidising modernisation and improvement of the equipment, nor by providing scientific research to solve the problems met in intensive farming. This is at least partly to be ascribed to the peculiar organization of fisheries research in Italy, largely under the auspices of universities, and therefore not necessarily directed towards the needs of the fishing industry.

F. GENERAL IMPRESSION

Mullet farming Italian-style has much in common with the system of mullet farming practised in Israel. There are, however, also noteworthy differences: in Israel mullet is farmed in slightly brackish water, which makes it possible to stock the ponds both with carps and with mullet, a combination which has certain advantages. In Italy the salinities in the valli are much too high for such a type of mixed farming. On the other hand, Italians are certainly not so keen on carps as the Israelis.

As far as labour is concerned there is a striking resemblance: the director of the Val Pisani can draw on the assistance of hands working in his agriculture enterprises in time of heavy work load, whereas the leader of the Israelian fish farm, under such circumstances, asks for hands normally working in other sections of the kibbutz. A striking difference is that the Italian fish farmer does not spread manure or inorganic fertilizer in his ponds to increase the production of fish food. The absence of carp, constantly churning up the bottom sediments, may have something to do with this. In any case, this way, blooms of undesirable organisms such as *Prymnesium* may be prevented. If one wishes to promote growth and fattening of fish it is always safer, though not always cheaper, to offer a type of food which can be directly digested and converted, than working along indirect pathways in an effort to promote the production of natural food, with the risk that a high percentage of the fertilizer falls on "weeds". Just as in animal husbandry on the land, it is the degree to which the natural processes can be kept under control, and thereby the economics of offering either fertilizer or feed, which guides the farmer. The case of the Val Pisani fish farm is a most interesting illustration of the decision farmers have to take in our time, with its increasing wages: what is the minimum production required to defray the annual costs of labour in a system of extensive culture, of which the crop fluctuates widely under the influence of a variety of natural factors? In case of the Italian "vallicultura", once flourishing with a minimum of labour and material, a noteworthy decline in the number of immigrating elvers and mullets, not to be compensated by collecting this fry in the coastal water in the neighbourhood, led to a serious diminution of the quantities of fish marketed. The situation was aggravated by the devastations brought about by a series of severe winters, and by the ectoparasite *Argulus giordanii.*

Simply waiting to see whether the situation will become better again, along natural pathways, is not a decision of our time. Considering the scarcity of fry and the availability of the fertile valli, a natural resource to be exploited rationally, the only reasonable decision is to take measures leading to a high survival of the still-available fry, of both eels and mullets. If there is not sufficient fry in the neighbourhood an effort should be made to buy it elsewhere. This decision taken, and considering the price to be paid for purchase of fry and for its transportation, the conclusion is reached that it is not economic to transfer it directly to a huge body of water such as a valle, with its multitude of predators, and other factors endangering the life of the young fish. They should therefore be reared in smaller basins and transferred step by step to larger ponds or basins as they grow and become less vulnerable.

This all requires considerable investment for building brood-houses and for making ponds, complete with canalisation, pumps, valves, etc., and entails expenses for feed and care.

It is a question of know-how and economics whether such a development will pay off. The fry may be expensive, but if a high and constant survival percentage can be counted on, these costs will not easily form a bottle-neck of the venture. For mullet and for *Sparus auratus* this stage has been reached, but for elvers the experimental phase is still in progress. The price for elvers (L.It. 5 000 to 10 000 per kg) may seem high and is caused by the demand from the consumers, but in reality this means that one single elver costs only about L.It. 4, very little when one can be sure of a reasonably high survival rate in the course of cultivation. The Japanese eel farmers have realized this, and are eager to buy European elvers for a much higher price than L.It. 4 per specimen.

If the first step has been taken towards intensive farming of a given species of fish, it should be realized that amortization of the invested capital, and expenses for feed and care require a given minimum price for the end product on the market. Successful farming on a large scale will moreover lead to a greater production, which involves the risk of a drop in the price. One of the first things to consider is the price paid for the same species caught in the wild, and the tonnage of those offered on the fish market. For fish like the mullet, *Dicentrarchus labrax* and *Sparus auratus*, there is hardly any risk from that side. The quantities caught in the wild are small and insufficient to cover the demand. As in trout farming, the cultivated product will soon rule the market.

Quite a number of problems of a biological, technical and economical nature have to be solved before mullet farming in the Alto Adriatico increases to ten times the present production, which seems a reasonable prospect. The most difficult problems are at both extremities of the production: procuring the fry and marketing the fish. The latter question could be solved by developing a good technique for deep-freezing individual mullet and by exploring the market for such a product all over Italy, to become independent of the local fish dealers and of the influence of fluctuations in the landings of sea fish. The former problem is the more difficult and intricate one: buying mullet fry from other sections of the Italian coast, or from the coastal waters of other Mediterranean countries, may seem a solution at short term, but involves, in the long-run, the risk of a declining production in such areas too. It seems of paramount importance to organize thorough research on the factors affecting the natural reproduction of mullet and the migration of the fry towards the coast. A good insight might perhaps lead to improvement in the situation. However, one should not count too much on such a solution, and so forget to strive towards another way out: artificial rearing of the fry from eggs taken from mature females. In principle this must be possible, but the difficulty lies in how to obtain adults ready to spawn, and how to create optimal conditions for rearing the fertilized eggs to a stage ready to enter the estuaries. Much work has already been carried out, especially in Israel and Hawaii, but thus far the rearing procedure is not yet ready for application on a technical scale. The "Centro Ittiologico Valli Venete" is carrying out research in this field and has already obtained some laboratory results with *Dicentrarchus* and *Atherina*. One should, however, consider that the ideal site for such work is to be found under hydrographic conditions identical with those prevailing where natural reproduction takes place, i.e. in pure high salinity water, not to be found by the continental coast, rather on the

Fig. 27. Dr. G. Ravagnan enjoys a good-sized fried mullet grown in his own farm. (Photo P. Korringa.)

coast of some small islands. Investigations to identify the right type of food for the larvae and to produce that in large quantities are part of the research programme. If a system of artificial reproduction can be developed, the production of mullets, and also of highly priced specimens such as *Sparus auratus* and *Dicentrarchus labrax*, can be raised to an unprecedented level, and could eventually serve the market in a very large area.

The rapid switch-over from extensive to intensive farming in the azienda Val Pisani has only been possible since the director-proprietor Dr. G. Ravagnan could, as a wealthy industrialist, decide to invest large sums of money, to be amortized in the long-run, without feeling the need to defray the initial expenses from the revenues from sale of fish at this moment. Considering the risks of this venture, it is therefore all the more surprising that Dr. Ravagnan does not try to monopolize his new system of farming, but divulges the acquired experience to all the participants of the C.I.V.V., and sees as ideal, a wholesale modernization of the valli cultures of the Alto Adriatico, leading to a dramatic increase in the production of high-quality fish for the Italian market.

G. BIBLIOGRAPHY

Anonymous, 1971. *Argulus giordanii*: note sul ciclo biologico. Bollettino informativo per i soci C.I.V.V., Ca Venier, pp. 13—25.

Anonymous, 1972. The Grey Mullet. Induced Breeding and Larval Rearing. Oceanic Institute Waimanalo, Hawaii, Rep. No. 01-72-76-1, 202 pp.

Anonymous, 1971. Notizie su un allevamento sperimentale di *Dicentrarchus labrax* (Branzino o Spigola) con alimenti secci. Bollettino informativo per i soci C.I.V.V., Ca Venier, pp. 93—101.

Arcarese, G., 1970. Primi tentativi di reiproduzione artificiale del latterino (*Atherina boyeri*). Riv. Ital. Piscic. Ittiopathol. A.V., 2: 27—28.

Arcarese, G., Ravagnan, G. and Ghittino, P., 1972. Primi risultati positivi di fecondazione artificiale nel branzino (*Dicentrarchus labrax*) su vasta scale. Riv. Ital. Piscic. Ittiopathol. A.VII, 2: 27—33.

Brian, A., 1959. Descrizione di un nuovo Argulus (*Copepode branchiuro*) della Laguna di Venezia. Tial, Genova.

Bullo, G.S., 1891. Piscicultura Marina. Stima delle coltivazioni in acqua salsa, Padova, Stabilimento Prosperini.

Bullo, G., 1940. Le valli salse da pesca e la vallicultura. La Laguna di Venezia, Vol. III, P. VI T XI. Off. Graf Carlo Ferriari, 18 pp.

Ghittino, P., 1969. Piscicoltura e ittiopatologia: Vol. I, Piscicoltura. Edizione Rivista di Zootecnia, Torino.

Ghittino, P., 1970. Piscicoltura e ittiopatologia: Vol. II, Ittiopatologia. Edizone Rivista di Zootecnia, Torino.

Ravagnan, G., 1972. Struttura e organizzazione di un moderno e grande impianto per l'allevamento di pesce bianco di mare e di anguille. Unpublished report, 250 pp.

Chapter 3

FARMING THE MILKFISH (CHANOS CHANOS) IN INDONESIA

Example
Milkfish farm "Gembira Bati", Pelelangan Ikan, Kali Baru, Tanjung Periok, Java, Indonesia.
Proprietor: Mr. Chalil Fadil, Jalan Jambi Lorong 44, No. 4, Tanjung Periok.
Period of observation: 1971—1972.

A. BACKGROUND

A.1. General principles

It is not known precisely when milkfish farming started in Indonesia, but from the Javanese law code "Kutara Menana" it can be deduced that it dates back to at least 1400, at the close of the Hindu period. Possibly experience gained in the salines of the Isle of Madura, where convicts had to work, may have induced its development. Schuster, an authority in this field, said that this ancient industry was "based on technical skill in pond construction and an almost instinctive understanding of the physiological requirements of the cultivated fish". Whatever the historical development may have been, it is clear that the milkfish lends itself admirably to farming because of its euryhalinity, its great resistance to diseases, its ability to withstand low oxygen levels, and the wide range of food organisms it can thrive on. The most important characteristic is no doubt its herbivorous nature: feeding predominantly on the organisms of the first link of the food chain, it can be grown in dense populations in adequately constructed ponds without administration of special food.

The milkfish is a marine fish of katadromous nature. It reproduces in the sea, but its larvae migrate to estuaries and brackish marshes in search of their favourite food: diatoms and blue-green algae. The young fish will grow rapidly there, and may stay for years in brackish or almost fresh water before they turn to the sea again for sexual maturation and spawning. If shallow ponds are constructed in the coastal area, within reach of the tides, milkfish larvae may enter there to stay. As a rule, the numbers entering freely are too small to base a farming procedure on, so the general practice is to fish especially for the larvae when they approach the coastline and to transfer the young fry to the ponds. Fry are not always found in abundance in the area were the ponds are constructed, but it is technically quite easy to transport the juvenile fish there, even over considerable distances. The milkfish farms of western Java get most of their fry from far-away beaches. Local people catch the milkfish larvae and it is via one or more middlemen that they finally reach the farm. An important feature in milkfish farming is adequate preparation of the ponds before the introduction of the young fish. The ponds must be as free as possible of predators, and next, the development of the favourite food organisms of the milkfish must be promoted. It is general practice to put the larvae in small and shallow ponds, and to transfer them, in due course, to larger and somewhat deeper sections of the fish farm. Though the fish can be kept growing for several years, which makes them reach quite a large size, it is common practice to keep them less than a year and to market them when they have reached a weight of 350 to 500 g each.

A.2. Biology of the species

The milkfish *Chanos chanos* (Forskal) (Fig. 28), locally called "bandeng", belongs to the Clupeoid fishes. It is an elongate, moderately compressed fish with silvery scales. Dorsal and anal fin have a broad basal sheath. The former is elongated posteriorly. The caudal fin is very large and deeply forked. The milkfish is glossy blue to bluish-olive above, the top of the scaleless head and the dorsal fin show a yellowish-olive hue, other parts of the fish are brilliant silver all over. Its mouth is small, terminally placed, non-protrusible and without teeth. The milkfish has numerous exceedingly fine gillrakers, closely set in two diverging rows. Behind the gill cavity are the accessory branchial organs. The intestinal tract is extremely long with many convolutions and numerous pyloric appendages.

Adult milkfish, living in the sea, reach a size of well over 100 cm, and a weight up to 15 kg. The milkfish has a wide geographical range, covering virtually the whole tropical and subtropical sectors of Pacific and Indian Oceans, i.e. from 40° East to 100° West, from 30 to 40° North to 30 to 40° South. It requires a minimum water temperature of 15°C, becoming para-

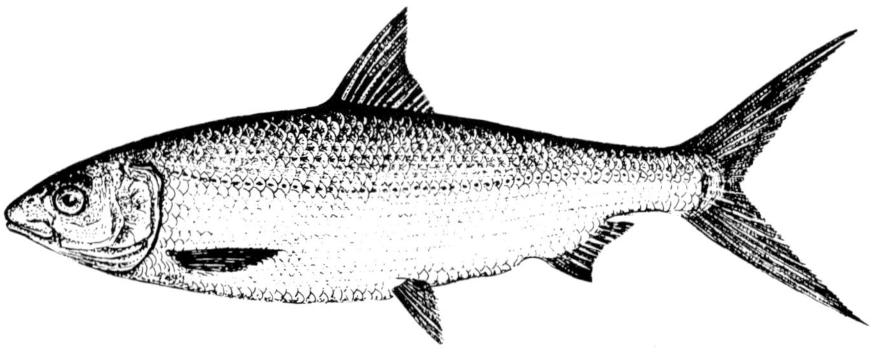

Fig. 28. The milkfish, *Chanos chanos* (Forskal).

lysed at temperatures as low as 13° C. Therefore it occurs only seasonally in the northern and southern fringe of its range. The milkfish is a marine species with pronounced katadromous features. It is rarely caught in the high seas. Spawning takes place some 5 to 15 miles offshore, in clear, rather shallow water (up to 20 or 30 m deep), above a bottom of sand or corals. A large female spawns millions of eggs. On Java's north coast there are two distinct spawning seasons, one in April—May, the other in September—November.

Shortly after hatching, the larvae, carrying a yolksac for about 2.5 days, start to migrate towards the coast, where they arrive when they are 10 to 15 days old. They may sojourn in shallow coastal water for about 4 weeks, and search in marshes and estuaries for brackish water rich in their favourite food: diatoms and blue-green algae. They grow fast, reaching 50 to 70 mm in about one month after arrival at the coast, 120—150 mm in 2 months, and some 400 mm (with a weight of 250—500 g) when they are 8 to 10 months old. They can stay several years in low salinity water, reaching a weight of 2 500—3 500 g in 4 years, but will show no gonad maturation there. Only after migrating back to the sea does gonad development start, followed in due course by spawning.

The milkfish larvae feed principally on diatoms, but gradually blue-green algae, among which species of *Oscillatoria* and *Phormidium* prevail, become their favourite food. Live and elastic algal threads are eaten in modest quantities at a time, those somewhat softened by decay more eagerly. From a size of 35 cm upwards, the milkfish can also feed on green algae and even on rooted plants such as *Ruppia* and on partly decomposed Characea, but blue-green algae remain the favourite food.

The fine gillrakers and the long intestinal tract make it possible to collect and digest the kind of food taken.

A.3. Geographical situation

Milkfish farms are found on several sections of Java's northcoast; large acreages of ponds are situated north and south of the city of Surabaja, and further bandeng farming is performed on the Isle of Madura, and also locally elsewhere in the Indonesian archipelago.

The milkfish farm "Gembira Bati" is situated on Java's northcoast, 16 km northeast of Tanjung Periok, Jakarta's seaport, near the village Pantai Bakti in the sub-district Tjabang Bungin of Bekasi. The ponds are constructed on the left bank of the southernmost branch of the Tjitarum, some 500 m from the coast. The exact geographical position is 6° 01′15″ S and 107° 00′ E. The farm covers 60 ha of which 45 ha are water.

A.4. Hydrographic pattern

The tambaks, the ponds for milkfish farming, are not all of the same hydrographic pattern. The milkfish is very euryhaline and can be farmed in

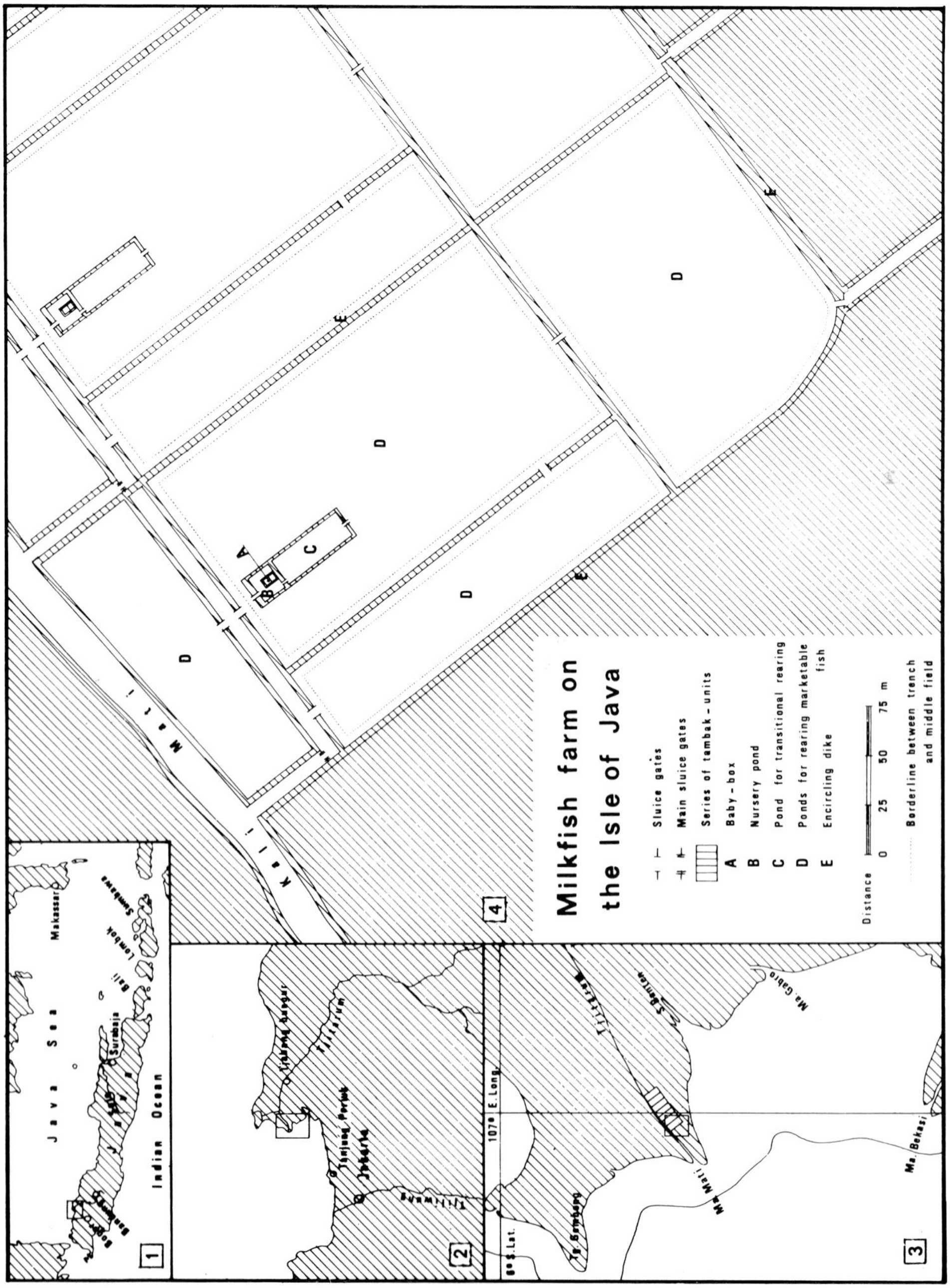

Milkfish farm on the Isle of Java
Sluice gates
Main sluice gates
Series of tambak - units
A Baby - box
B Nursery pond
C Pond for transitional rearing
D Ponds for rearing marketable fish
E Encircling dike
Distance 0 25 50 75 m
Borderline between trench and middle field
Kali Mati
Java Sea
Indian Ocean

water of high salinity, but also in brackish and almost fresh water. One distinguishes three categories of ponds:

(1) Tambaks receiving water predominantly from the nearby sea. The salinity fluctuates between 25 and 35‰.

(2) Tambaks situated at river estuaries under the influence of the tides. The water in the ponds is brackish, the salinity, varying with the season, being rather high in the dry months.

(3) Tambaks situated further inland, receiving brackish water via long channels at high tides only. The salinity is low, sometimes less than 5‰, and hence the vegetation in the ponds closely resembles that of fresh water communities.

The Gembira Bati farm belongs to the second category, receiving its water from the Kali Mati, the southernmost branch of the Tjitarum, close to the site where it debouches in the sea, and also directly from some channels leading to the sea.

Tides. The coastal area is very flat, so that the ponds can be filled and flushed on the tides, even though the tidal range is quite modest here. This section of the northcoast of Western Java shows a diurnal tide with a range up to 110 cm at spring tide and 40 cm or even less at neaps. The ponds are made in such a way that their bottom attains a level some 40 cm beneath mean high-water spring. The ponds are provided with sluice gates, of simple construction, which makes control of the water level possible.

Water temperature. Since the ponds are shallow, the water temperature is strongly under the influence of weather conditions. The temperature in the ponds rarely drops below 24° C, and may occasionally rise to as high as 38° C. Since the latter is really too high, the ponds should be flushed in periods of spring tide when the weather is extremely warm.

Salinity. The ponds of the category to which the Gembira Bati farm belongs are filled at high tides with estuarine water. Therefore the ponds are brackish, with a salinity ranging from 15 to 35‰. Evaporation may lead to further increase of the salinity. The milkfish itself will not be killed by high salinities but the vegetation on which it thrives may suffer considerably. Therefore, it is strongly advisable to keep the salinity under control and to flush the ponds with estuarine water often enough to prevent the building up of too high a salinity. But the farmers are dependent on spring tides since there are no pumps available.

Depth of the ponds. The vegetation on which the milkfish feeds develops most abundantly at a water depth of only 10 to 15 cm. For the juvenile fish, very shallow ponds may be used (10 cm water depth), but older fish stay away from such shallow areas, which presumably has something to do with their way of feeding on the bottom vegetation — in an almost vertical position. Therefore the pond should be everywhere at least 30 cm, with the exception of the nursery pond. It is not advisable to make the ponds too deep, since that would greatly reduce the quality and quantity of the bottom vegetation.

Nature of the bottom. Since the ponds of the category under consideration are constructed in the mangrove swamps fringing the coast, the bottom consist of soft mud. As a rule this is a layer of brownish black hydrophylic mud, biologically active and containing about 4% of humus plus large amounts of clay. Such soils form an excellent environment for the development of blue-green algae and discourage the growth of green algae.

Chemical composition of the water. Estuarine waters, containing a fair percentage of river water, have a brownish colour, because of their laterite content. The pH of the water fluctuates usually between 7.1 and 7.9. The oxygen content varies widely in the ponds, since the dense vegetation shows an intense photosynthesis during daylight, easily leading to oversaturation with oxygen. Oxygen bubbles develop in the vegetation mat and large patches of it may thereby be buoyed up, to descend again at dusk. The milkfish can survive in water with remarkably low oxygen values, but great care is required when poisonous hydrogen sulphide tends to develop. This is indicated by development of white patches on the bottom of the pond. Prompt action is then required to avoid serious losses in the fish stock. Flushing is again the best remedy.

Fertilization. There are several ways of increasing the vegetation which serves as daily food for the milkfish. Before stocking the ponds with fish, the farmer may administer cow dung or some 200 kg of urea per hectare, but when the ponds are stocked he can only utilize vegetable manure. In the Gembira Baṭi farm vegetable fertilizer is used exclusively after stocking the ponds, some 4 000 to 5 000 kg/ha/year: leaves of the mangroves *Avicennia* and *Rhizophora* being abundantly available in the area, as well as branches of the aromatic shrub *Pluchea indica* (Compositae), and sometimes even rice straw. This vegetable material, deposited periodically in the shallow central part of the ponds, decays readily and stimulates the growth of blue-green algae. Manuring requires considerable caution: oxygen depletion easily leads to production of hydrogen sulphide gas.

Pond vegetation. On the muddy bottom of the shallow ponds a real mat of vegetation may develop. Blue-green algae, predominantly belonging to the genera *Oscillatoria* and *Phormidium* form the backbone of the mat, but diatoms such as *Pleurosigma's*, *Gyrosigma's* and *Nitzschia's* abound in it. A variety of protozoans and rotifers, together with small crustacea and insect larvae will also be found in it upon closer inspection. This vegetation mat, of which patches come to the surface in periods of intense photosynthesis, is locally called "tai-air" (water dung). It is on this mixed vegetation that milkfish in all developmental stages feed eagerly. Sucking and smacking noises are heard when the fish feed from the floating patches. Occasionally green filamentous algae, Characea and higher plants such as *Ruppia* will be found in the ponds, but these can only serve as food for the larger specimens of *Chanos chanos.*

The development of the desired type of vegetation can be furthered in a virginal pond by transferring patches of the mat of benthic algae from an already productive pond. Occasionally the food supply in the ponds is increased by administering green algae or the red weed *Gracillaria* collected at the sea shore.

A.5. Legal aspects

The organisation of the "Gembira Bati" fish farm, founded in 1960, and now existing of seven units, is quite complicated, but represents the normal pattern for the area under consideration. The owner, Mr. Chalil Fadil, of Arabian descent, lives in Tanjung Periok, 16 km away, as the crow flies. He has to pay the rent (Rp. 1 000/ha/year for the gross acreage in use) to the owner of the land, the Forestry Service of the Government. Further, all the investments, such as construction of new ponds and sluice gates, are for his account. Larger maintenance works such as drastic repair of sluice gates and periodical deepening of ponds and canals are among his duties. He also purchases the necessary fishing gear and utensils. It is the owner who purchases the fry, but this latter item is put on the expense account, to be subtracted from the gross income, before sharing the profit.

The Gembira Bati fish farm is run by seven participants, one for every unit, who live with their families on the spot. The participant-caretakers survey the ponds, canals and sluice gates, and carry out smaller repair works. They are responsible for the true farming: for administering vegetable manure; for surveying the quality of the water; for flushing the ponds, whenever necessary; for looking out for possible leakages of the ponds, screens and sluice gates; for general surveillance against theft and poaching; and for controlling predators and competitors. The participants get a share (one-third) in the net profit of their units.

The assistance of odd hands is called in for the heavier work, such as construction of new ponds or periodical deepening of the existing ponds and canals, but also when ponds are to be fished out. In the former case the expenses are for the owner. For assistance in fishing out ponds the odd hands get a payment in kind, i.e. a given percentage of the catch.

B. SEQUENCE OF OPERATIONS

B.1. Construction of ponds

The Gembira Bati fish farm started in the year 1960 with one unit, measuring 8 ha. A unit is built up of rectangular ponds of different dimensions, among which, is a small and shallow nursery pond, measuring from 500 to 1 000 m^2. In the latter, one usually finds the so-called baby-box, a tiny section measuring a few m^2 only, to be used for the first few days after arrival of the fry. About 10% of the acreage of each unit is covered by a pond for transitional farming. The remainder consists of large ponds for growing marketable fish. The sections of the units are interconnected with sluice gates provided with bamboo screens, to allow flushing without loosing fish and without intrusion of predators. The entire unit gets its water via a deeper canal leading to the river; normally closed off with a sluice gate of more sturdy construction.

The creation of a tambak-unit in the muddy alluvial grounds of the mangrove belt is not an easy matter. First, most of the coarse vegetation has to be eliminated, mangrove trees included. The next step is to excavate a trench, 3 to 4 m wide, 0.80 m deep, to encircle the whole unit. The excavated material is used to construct an encircling dike outside the trench, the crown of which should stand about 50 cm above high water spring level. In later phases, this dike should be kept under constant control and repair, for the material of which it is constructed does not always prevent leakage, and the soft subsoil it is often built on, cannot always carry its weight. When trench and dike are ready, it is time to install the main sluice gate. Because of the soft subsoil in this area, it is not built of concrete or masonry, but rather of meranti wood from Borneo. When the water level in the unit can be adequately controlled, the subdivisions in the unit are made with lower dikes; excavated where necessary, the dug-out material is transported on bamboo rafts to the sites where dikes are to be constructed. The bottom of the middle fields of the main rearing ponds should lie at a level of about 40 cm beneath mean high-water spring, so that they can get a water layer of about 35 cm; the encircling trench should be about 80 cm deep. The bottom of all the ponds should be levelled carefully and all tree stumps should be eliminated. The slope of the dikes should be steep to discourage the settlement of rooted plants. Mangrove trees (*Rhizophora*) are usually planted on the dikes to make them stronger and to offer some shade for the fish in the ponds.

Since construction of a tambak unit is carried out by hand, special spades and mattocks being used, among which, a narrow-bladed spade to dig in clayish soil, it usually takes a long time, often years. Temporary shortage of money may lead to prolonged interruption of the work. Moreover, the unit cannot be used immediately for stocking with milkfish, for in the first years

Fig. 29. Tambaks, ponds for milkfish farming, close to the sea in the vicinity of Jakarta, Java. (Photo courtesy W.H. Schuster.)

there is leakage in the dikes and sluices, especially when they are constructed on a soft subsoil. Therefore a unit is first used for growing shrimps and wild-fish, either coming in freely from the estuary, or stocked especially for the purpose. As a rule one stocks the units with milkfish fry for the first time 3 to 4 years after its construction.

B.2. Preparing a pond for stocking

When a pond has been used for growing milkfish it should not be stocked again immediately after fishing out. The pond should be drained and exposed to air and sun sufficiently long to make the soil surface crack (Fig. 30). Such a procedure is considered of paramount importance, since it kills a variety of predators, and speeds up the mineralisation processes in the top sediment layers, thus improving its structure. Whenever necessary, the bottom of the ponds should be tilled, raked and levelled, and in this, the greatest care is always bestowed on the shallow nursery pond. Drainage ditches and ring channels should be kept at the appropriate depth and efficiently freed from all predatory fish and crustaceans. Sometimes the fish *Piscodonophis boro* (Ham. & Buch.) and other dangerous predators emerge alive from the dried out soil when some water is let into the pond again. Then the ponds should

Fig. 30. Before stocking, the bottom of the pond should be thoroughly exposed to air and sun. (Photo courtesy W.H. Schuster.)

be drained again and deeper sites should be carefully fished out. Reduced losses among the stocked milkfish will amply pay for all the care taken in preparing the ponds. Nursery ponds get a water layer of 10 cm; the ponds for the large milkfish 30 to 40 cm. The ponds are always filled gradually, first a layer of 3 to 5 cm for 10—12 days, and in about 15 days a good mat of blue-green algae has developed on the bottom. The whole preparation of a pond, the period that it has to lie fallow for thorough reconstitution, takes some 30 to 50 days, which means that every pond can only be used for growing fish for 10 months per year.

B.3. Collecting the fry

The milkfish farmers of Western Java have to purchase the fry from areas where it is amply available. Northeastern Java, and especially the isle of Madura, are the traditional regions from which considerable quantities of milkfish fry are obtained. Transportation over hundreds of kilometres is technically possible. The milkfish farmers do not fish the fry themselves but buy it via a series of peddlers, dealers and middlemen, from local people. When the milkfish fry approaches the coast, thousands of people try to get hold of them, always on gently sloping sandy beaches with clear water. The usual technique is to operate with a lure line ("blabar"), a rope of 15 to 30 m long into which leaves of palm trees or sugar-cane, or of alang-alang grass (*Imperata arundinacea*) have been plaited to reach a diameter of 10 to 20 cm. Such a garland is either stretched between two bamboo poles, perpendicular to the coastline, or paid out spirally from one pole. The young fry, locally (Madura) called "nèner", seek shelter under and in between the leaves. The fishermen, wading out in the sea, come with a triangular scoopnet of very simple construction to catch it.

The fry is collected in 2—3 l unglazed earthenware pots, filled with a mixture of 10 parts fresh water and 1 part sea water, to reach a salinity of 3 to 4‰. Milkfish fry can stand the sudden transition in salinity, but larvae of mysids and several other unwanted organisms are killed this way and can then easily be eliminated. The autumn run of milkfish larvae is considered the best, and peak catches can be expected round about the full and new moon spring tides. The pots can hold up to 1 000 specimens. Soon the fry is transferred to baskets called "waluh", woven of thin strips of rattan, made watertight with an exterior coating of cement and an interior lining of wax or pitch. The flat round baskets, usually with a diameter of about 60 cm, contain 15 l of water, in which 20 000 to 40 000 newly caught milkfish fry, measuring 11 to 13 mm in length, can be put. Special scoops, made of a large shell, are used to transfer the young fry from the pots to the baskets, and each time they are transferred to another container, unwanted fry of other fish, such as species of *Mugil*, *Megalops*, *Elops*, *Dorosoma* and *Stolephorus* are eliminated. A keen eye is needed to identify the milkfish larvae on basis of their special pigmentation and characteristic swimming behaviour, but the potential predator *Elops hawaiiensis* is rarely completely eliminated.

Fig. 31. Catching milkfish fry with the aid of a lure line.

Fig. 32. On the isle of Java, fry of freshwater fish are transported in baskets called "karamba", almost identical with the "waluh's". The difference is that a waluh is constructed of rattan strips, is made watertight with cement and has a round bottom, whereas the karamba is made of bamboo strips, made watertight with pitch and has a more or less square bottom. The way they are carried is exactly the same. The photo displays a carp fry transport in Tjiandjur. (Photo P. Korringa.)

B.4. Transporting bandeng fry

If the "waluh", the round transportation basket, is filled with a thin layer of water (3 to 4 cm above the slightly elevated centre of the bottom), the bandeng fry can be transported over considerable distances by road or railway, without aeration, provided the utmost care is taken to renew the water every day and to feed the tiny fishes adequately. For transportation over hundreds of kilometres, not more than 10 000 fishes can be put in each basket. When no sea water is available on the road, some crude sea-salt should be added to fresh water to give a salinity of 2 to 3‰, judged by taste. The fry are fed with rice-flour, often slightly roasted, or minced hard-boiled egg-yolk. A teaspoon of flour twice a day is sufficient for the 10 000 juvenile fishes contained in a "waluh". As long as all the fish in the basket swim school-wise in the same direction, there is no need to worry. If not, the water must be renewed immediately. A whole series of dealers and middlemen passes on the fry, and at each transfer the fishes are counted on basis of a sample. Still contained in the practical carrying baskets, which can be stacked on top of each other, they finally reach the farmer. The "Gembira Bati" fish farm purchases about one-third of its fry in May—June, and about two-thirds in November—February. It is not always easy to obtain the required quantities from Eastern Java and Madurã, but in recent years notable quantities could be purchased from Makassar and Lombok, and transported by plane. The buyer should have a keen eye for the quality of the fry he purchases. Its quality depends on the time of the year and the part of the run from which the fry is caught, and above all, on expert care during storage and transportation.

B.5. Stocking the nursery ponds

When the bandeng fry arrives at the farm, where nursery pond and "baby-box" (Fig. 33) are duly prepared, it is advisable to check whether there is a dangerous difference in salinity. Though the fry can easily cope with a rapid

Fig. 33. A "baby-box" to which the newly arrived milkfish fry first transferred. (Photo courtesy W.H. Schuster.)

transition from saline to brackish water, the reverse requires some caution. The salinity should never be too high in the nursery pond (if possible, less than 30‰), and the transition should be a gradual one. Early morning or late evening is preferred for bringing the fry to the "baby-box". In this small pond one can observe their behaviour, and when everything seems to be all right the earth partition between "baby-box" and nursery pond is cut. Then the fry can spread over the shallow nursery to feed on the rich food mat which has developed there. The nursery ponds are stocked with some 30 of the tiny fishes per m^2 (i.e. 300 000 per ha). This density is really quite high. Schuster (1960) tells us that optimal growth can be expected at a stocking rate of only 10 fishes per m^2. Stocked at a density of 30 to 50 per m^2, one may expect them to reach a size of 20 mm at the end of the first week, 29 mm after the second, 37 mm after the third, 42 mm after the fourth and 53 mm after the eighth week. Roughly double that rate is achieved at a stocking density of $10/m^2$.

Fig. 34. A nursery pond, here centrally placed in a larger pond. After cutting the dike the juvenile fishes can swarm out into it. (Photo courtesy W.H. Schuster.)

B.6. Transferring fingerlings to the larger ponds

A farming unit of the usual size, of 8 to 10 ha, has only a very small percentage of its acreage in the form of nursery pond, further, about 10% of the ponds are kept for transitional rearing, and nearly 90% of ponds are designated to produce marketable fish. The larger the fish, the more space and food they need, but the smaller the pond section they are kept in, the easier is control and maintenance. There are sluices which connect the various ponds of the tambak unit and there is, as a rule, no need for fishing out the ponds containing young fish, for the experienced fish farmer knows exactly how to transfer his fish to the desired tambak section, making use of the rheotaxic behaviour of the milkfish. If water is let in about spring tide, the fish get very excited and tend to concentrate at the bamboo screens through which the water comes in. This behaviour makes it possible to guide them to the desired section, or eventually to catch them near the sluice gate to transfer them in baskets or kerosine tins to some far away section of the farm.

In the shallow nursery pond the young fishes sojourn for a period of 40 to 50 days. When stocked at a density of 30 per m^2, one may expect them to grow to a size of 50 mm in that period, with a weight of about 1.5 g each. The losses suffered in that period are estimated at 30 to 50%. Under especially good care, this figure can be lowered to 20%.

After leaving the nursery pond as fingerlings of about 50 mm length, the young fishes come into a much larger and deeper transitional pond, in a density of 15 000 to 20 000 per ha, from which they are spread later over the spacious rearing ponds of the whole unit. Finally they are grown in a density of 1 000 to 1 500 per ha in the larger compartments of the unit.

B.7. Harvesting the marketable milkfish

After a total rearing period of 8 to 10 months (viz. 1.5 months in the nursery pond, 2.5 months in the transitional pond, and 4—6 months in the large growing ponds), the milkfish may have reached a weight of 200 to 350 g each, which is considered to be a good size for the market. When it is decided to harvest the fish, the farmer gets the assistance of a fair number of casual workers, who, armed with cast nets, operate systematically by wading across the pond in a long row (Fig. 35). They chase most of the fishes to the deeper site at the sluice gate, where they are easily caught. The entire crop cannot be harvested in one passage; one or two more passes are needed. When the tide allows, the pond can be drained, but even then one has to catch the fish at least partly with hand-cages and cast-nets in the deeper ring canal and drainage ditches of the ponds.

Though bandeng (milkfish) is the main product of the Gembira Bati fish farm, certain by-products are collected, among which number various species of shrimps and certain wild fish which enter the ponds — despite all precau-

Fig. 35. Fishing off a milkfish pond with cast nets. (Photo courtesy W.H. Schuster.)

tions taken. Shrimps are caught in special bamboo fyke nets throughout the year. Since the total rearing time for milkfish lasts from 8 to 10 months, of which only 4—6 months are spent in the large growing ponds, covering 90% of the acreage, the farming units can yield two crops of marketable fish per year. This is very efficient, since one can thus take advantage of both natural spawning peaks of the bandeng by alternatively stocking the ponds with east monsoon fry ("labuh") and west monsoon fry ("kapat").

B.8. Marketing the milkfish

The fish harvested from the ponds of the Gembira Bati farm are quickly transported to the fish auction of Tanjung Periok, some 16 km away. One transports the fish over sea by a motor-boat, especially hired for the purpose. The bandeng is greatly appreciated on the Isle of Java, and sells at a good price. The bulk is sold in the fresh state, packed with ice. Where no ice is available, one often makes a special, less perishable product by boiling the milkfish in brine and adding some Curcuma root, which gives the fish a

Fig. 36. Packing milkfish with ice in baskets lined with banana leaves. Sembajat, eastern Java. (Photo P. Korringa.)

Fig. 37. Milkfish packed with ice in baskets lined with banana leaves. Sembajat, Eastern Java. (Photo P. Korringa.)

Fig. 38. Baskets with fair-sized milkfish arrive on the market. (Photo courtesy W.H. Schuster.)

special flavour. Hot-smoked milkfish is a real delicacy, but cannot be kept long in a tropical climate unless deep-frozen. Small milkfish, canned in tomato sauce, is a very good product too.

C. FARMING RISKS

C.1. Hydrographic conditions

Flooding of the river, by which the tambak has been constructed, is one of the greatest risks the bandeng farmer is faced with. If the dikes and sluices are not high and strong enough, all the fish from the ponds, in which so much money and labour have been invested, may disappear overnight. Increasing deforestation in the hinterland may lead to more frequent and more serious floods.

High salinities, low oxygen levels, and above all formation of the poisonous hydrogen sulphide may endanger the life of the fish and their vegetable food. It is the farmer's duty to control the quality of the water in his ponds and to flush them whenever necessary. On the other hand, the farmer has to administer vegetable manure, which requires much oxygen for its biodegradation. Especially at neap tide, when flushing is virtually impossible, he should be very careful, for not only bacteria and other micro-organisms destroying the dead vegetable matter consume a considerable amount of oxygen, but also the living algae, which produce so much oxygen during the daylight hours, consume an appreciable quantity of oxygen during the dark. This easily leads to very low oxygen values towards dawn. As long as there is an appreciable quantity of oxygen, hydrogen sulphide will not be formed in the water.

Weather conditions may counteract the efforts of the farmer, leading to high salinities in periods of drought and to insufficient drying out of the bottom of the ponds in the period they have to lie fallow in periods of excessive rain.

The farmer should develop a keen eye for aberrations in water quality, for they have neither aerometers nor chemical reagents at their disposal to check salinity, oxygen and hydrogen sulphide levels.

C.2. Predators

Predators are a serious menace for the milkfish grown in the ponds, especially for the juvenile specimens in the nursery ponds. Already during transportation of the fry, the large-mouthed larvae of the clupeoid fish *Megalops cyprinoides* prey upon the milkfish fry. The water bathing the shores of the Isle of Java are commonly populated with a variety of predatory species, and every effort should be made to keep those out of the ponds. Young predacious fishes do, however, manage to come into the ponds with the sea water entering the gate during filling. Perfect screening of the gate is an extremely difficult matter and some fish make their way through small leaks in the dikes. A few predators of 150 mm length in a nursery pond may play havoc among the bandeng fry. Serious predators, preying on juvenile milkfish up to a length of some 100 mm are the fish *Lates calcarifer*, *Elops hawaiiensis*, *Chirocentrus dorab*, *Sphyraena jello*, *Polynemus heptadactylus*, *Ophiocephalus striatus* and *Piscodonophis boro*. A fish like *Lates calcarifer* has an excellent market quality itself, but if one considers that it must have devoured 6 to 8 times its weight in milkfish and shrimps, it is clear that no bandeng farmer will enjoy its presence in the ponds. The water snakes *Cerberus rhynchops* and *Fordonia leucobalia*, and the frog *Rana vittigera* are also to be considered as serious predators in the bandeng farm. Further there are birds such as kingfishers, terns, herons, cormorants and even eagles which like to feed on the young milkfish.

The losses inflicted by all these predators are so serious that the utmost exertion to control them is an absolute necessity in milkfish farming. During the period the ponds lie fallow the vast majority of the predator population will disappear. For the deeper sections one needs intensive fishing out, and, whenever possible, application of a fish poison like endrin (0.2 l/ha) to kill the last ones. Perfect maintenance of the dikes, avoiding all leakage, and of the screens, is a sheer necessity. Snakes and frogs must continuously be captured, trapped and killed; birds are chased away with shot-guns, fireworks and other contraptions.

One could think of stringing nylon wire across the ponds to keep off the birds, but this would be cumbersome during fishing out. Still, this method is practised in tambaks elsewhere on Java. To keep out young specimens of predatory fish, rotating plankton sieves could be installed in the sluice gates, but these would be rather expensive, and moreover, require electricity or another type of energy to keep them in motion during filling or flushing of the pond.

C.3. Parasites and diseases

There are remarkably few parasites and diseases which endanger the life of the milkfish in the ponds. Schuster (1952) ascribed this fortunate phenomenon to the continual and complete renovation of the stock, combined with the occasional occurrence of salinities of over 35‰, and to the system of periodical desiccation of the pond bottom.

C.4. Competitors

There are not very many creatures which feed by preference on the blue-green algae constituting the main food of the milkfish. As such, there are the mullet (*Mugil sp.*), which should therefore be considered as food competitors in the ponds. They often come into the ponds as tiny fish during the filling, since they show the same tendency as the young milkfish, to migrate towards the brackish water.

Gastropod snails belonging to the *Cerithidae* are serious competitors for food in the tambaks, especially the species *Centhidea cingulata*, *Centhidea djadjariensis* and *Centhidea alata*. One sometimes counts hundreds of them per square metre of pond, and they take not only a substantial part of the food destined for the milkfish, but also a noteworthy quantity of calcium for constructing their shells. Hand-picking is the usual remedy, but since this is a time-consuming procedure, the Governmental Fisheries Institute advises to utilize a molluscicide such as Brestan-60.

In ponds with a rather high salinity, Polychaete worms of the genus *Eunice* may be cumbersome, because their burrowing activites make the pond bottom too porous.

D. ECONOMIC ASPECTS

D.1. Rents

The rent to be paid annually for the gross acreage in use by the Gembira Bati milkfish farm amounts to Rp. 60 000.

D.2. Inventory items

Construction of a tambak requires a large capital investment. At the price levels of 1972, it would cost Rp. 400 000 per ha to construct a tambak. Purchase of a well established tambak of the category under consideration would require Rp. 500 000 per ha. A newly made tambak, a sheer framework, still requiring subdivision into sections, and a lot of auxiliary work and repair before it can be used for stocking milkfish, can sometimes be bought for only Rp. 100 000 per ha. A well constructed and properly maintained tambak can serve for milkfish farming for very many years in succession.

Items of major importance are the sluice gates, made of wood. A large meranti wood sluice, measuring 1 × 3 × 4 m, would cost Rp. 20 000, smaller ones, 0.50 × 3 × 4 m, Rp. 15 000. The life of a sluice gate is only 2 years. Apart from the ponds with their dikes, sluice gates and screens, there are few inventory items of importance. A shed is needed for storage of material, some spades, mattocks and nets. For fishing out a pond, the casual workers engaged usually come with their own castnets.

D.3. Expendable items

A major item, recurring every year, is the purchase of fry. The Gembira Bati fish farm needs from 100 000 to 200 000 bandeng fry per year, of which about two-thirds is bought from the autumn run. The price of the fry varies with its quality, with the quantity caught in the year under consideration, and with the transportation costs. One should count on a price of Rp. 3 000 to 4 000 per 1 000 specimens. The Gembira Bati farm does not buy or sell fingerlings, the juvenile fishes measuring 8 to 12 cm, which generally cost Rp. 20 per specimen. It is the owner of the pond who buys the fry and pays for it, but these costs are later deducted from the gross income of the farm. Among the other expenses are the costs for manure to fertilize the ponds. Collecting, transporting and scattering the vegetable manure will cost about Rp. 7 500 per ha. For pesticides to eradicate predatory fishes and snails, one should count on Rp. 250 per ha. Bamboo and fish screens cost Rp. 500 per ha per year.

It is necessary to deepen the ring canal every year. For this work the unit generally used is the "ukur": 4 m^2 surface by 0.20 m depth. Per "ukur" the worker receives Rp. 40. The annual deepening of the ring canals and tilling and levelling the bottom of the nursery pond costs about Rp. 4 000—5 000 per ha, to be paid by the owner in money, and to be carried out by casual labour.

D.4. Labour

The owner of the tambak complex Gembira Bati lives in Tanjung Periok and does not participate in the actual farming work himself. The seven participants-caretakers carry out the daily work: surveying the ponds, taking care of the general maintenance of the dikes, sluices and screens, manuring the ponds, controlling predators and competitors, and preventing theft. The members of their families will assist in these chores. The participants receive one-third of the net income of the farm as payment for their time and efforts. The remaining two-thirds is for the owner to compensate for his capital investment, annual maintenance expenses and risks. The odd hands, called in from the neighbourhood for fishing out a pond, which requires some 20 men-days per ha, receive payment in kind: 5% of the bandeng, and from 30 to 50% of the extraneous fish. They are assumed to bring their own castnets with them.

D.5. Yields

The Gembira Bati fish farm yielded recently almost 510 kg of fish and shrimp per ha per year, of which 300 kg was bandeng (milkfish), 80 kg *Tilapia*, 30 kg shrimps of various descriptions, and about 100 kg "wild fish". This is considered a reasonably good production, but there are tambaks which produce 1 000 kg per ha per year, attributable to stringent predator control.

D.6. Sales

Larger quantities of fish, especially of milkfish, are transported over sea to the fish auction in Tanjung Periok. Here one can count on an auction price of some Rp. 200/kg for bandeng (milkfish), Rp. 75/kg for *Tilapia* and Rp. 100/kg for the "wild" fish. For shrimps of larger size, one receives Rp. 450/kg, for smaller shrimps Rp. 80/kg.

The gross income of the Gembira Bati farm was, in the year under consideration, made up of Rp. 2 750 000 for milkfish, Rp. 270 000 for *Tilapia*, Rp. 1 012 500 for large shrimps, Rp. 180 000 for small shrimps and Rp. 225 000 for "wild" fish.

D.7. Additional expenses

For the transportation of fish to the market, a boat equipped with an outboard engine of 40 hp is hired. This costs Rp. 5 000 per day. For the labour involved in transporting the fish to the market, one pays 5% of the fish in kind. Auctioning the fish costs 5% over the total value plus 5% "Ipeda" tax (a contribution for regional construction).

E. GOVERNMENT ASSISTANCE

The Indonesian Government assists the tambak farmer in various ways. The rent charged for use of the ground is very low: Rp. 1 000/ha/year.

The Indonesian Institute for Fishery Research at Bogor (Lembaga Penelitian Perikanan Darat), operating with several field stations, carries out various research programmes in behalf of the milkfish industry. Through the mediation of the extension service it has given valuable advice on transportation systems for milkfish fry — air transport included, and experimented on both a laboratory and technical scale on fertilization schemes for fish ponds, including the proper way to administer urea in the pre-stocking phase. Control of predators and competitors, so important in the efforts to raise the production, is one of the topics of the fishery institute's programme. Advice on the proper use of pesticides is given. An extension service, whose participants are regularly informed about the results of the latest scientific achievements during special courses, brings the information right to the fish farms, and ensures a feedback of the problems encountered by the farmers themselves, to the institute.

One of the major problems of the milkfish industry is the increasing shortage of fry off the coasts of Eastern Java and Madura, where it used to abound. To supplement the quantities caught, collection and transport of bandeng fry has been arranged from far-away districts, such as Makassar, Lombok and Sumbawa, but it is feared that intensive fry fishing in those areas might, there too, in due course, lead to a decline of the catches.

It is not clear which factor is responsible for the shortage of fry. One often speaks of pollution, but since the bandeng reproduces well offshore, in clear water with a sandy bottom, it does not seem very probable that pollution is the main cause of the decline in the quantities of fry approaching the coast in the appropriate season. One wonders whether one should not think in terms of restricted local populations of bandeng, spawning year after year in the same limited area, and receiving too few recruits from the coastal water to keep their population at the desired level. Each female bandeng does spawn very many eggs, but the hazards are great for the tiny young fry, both in offshore and inshore water. It is conceivable that improved farming techniques, better sluices and screens, combined with a very intensive fishing for bandeng fry, gives too few specimens a chance to live a free life and to return to the sea in due course to mature and to reproduce.

In the short term, a way out of this difficulty is collecting fry further and further away, and bringing it in by plane, if necessary even from other countries, if it abounds there. In the longer term this may not help sufficiently. Considering the great number of eggs produced by each mature female fish, it would be interesting to rear the fry from the eggs. Thus far, all efforts to mature the bandeng in captivity with the aid of hormone injections have not led to practical results. It seems advisable to continue these efforts, but then by preference, in a site with clear high salinity offshore water with sandy bottom, since such are evidently the favourite conditions for spawning of the bandeng.

F. GENERAL IMPRESSION

Farming the bandeng in tambaks is an ancient industry, of which, the methods have changed but little down the ages. All the work is still carried out manually, and the tools used are of simple construction. Mechanization has not yet been introduced and it will not be easy to do so, especially not in an isolated area as that under consideration, where there are no transport roads and no electric power. An outboard engine, installed on the boat which brings the fish to the market, is virtually the only item of mechanization introduced in recent years.

Bandeng farming is a typical example of semi-culture: one leaves the production of fry fully to nature and the fish grow up on a natural crop of food, stimulated by administration of predominantly vegetable manure.

A production of 350 kg milkfish per ha in some 6 months * may not seem very much, surely not when compared with a vegetable crop, wet weight. From some other countries, one gets information of a higher yield per ha, but though it is possible to double the production by expert predator control and careful manuring, the very high figures encountered in literature (well over 1 000 kg/ha) are not comparable with those related to normal tambak operations. They should be explained in terms of a bio-industry in which food, grown elsewhere and specially prepared for the purpose, is offered to the fish in a pond, taking care of adequate oxygen content of the water. Such a bio-industry can yield a multiple production, calculated per unit of pond surface, but has a very different economical structure.

That production of 300 to 500 kg milkfish per hectare in half a year is, in reality, a high figure follows from comparison with the total fish catch in a very productive sea like the North Sea: 35 kg/ha/year, or of a very fertile shallow fresh water lake as the IJsselmeer (Netherlands): 50 kg/ha/year, of

---

* The total rearing period is 8—10 months, but only 4—6 months of those are spent in the large growing ponds covering 90% of the area, so that a tambak can produce two crops per year. A sophisticated stocking policy, keeping fishes of different size in one and the same pond and letting the smaller categories escape through the meshes of the cast-nets during harvesting, can still further increase the productivity of the tambaks.

which only 15 kg/ha is fit for human consumption. This difference should not, however, in the first place be ascribed to the efforts of the tambak farmer, administering vegetable manure to his ponds during the rearing process and if necessary urea in the pre-stocking phase, but rather to the fact that the bandeng is herbivorous. The milkfish feeds directly on the organisms of the first link of the food chain, chiefly on bluē-green algae, partly on diatoms, soft green algae and decaying rooted plants. That is the reason why one can attain such a high production per hectare. The fishes caught in the North Sea and in the IJsselmeer are carnivorous, feeding in their turn on herbivorous organisms, partly even on carnivorous species, and therefore belong to the third and fourth links of the food chain and not to the second as the bandeng. In general, one can claim that the total production per hectare declines about 90% for each time one switches over to organisms one link higher in the food chain. Therefore the quantity of milkfish produced per ha will have to be in the order of 10% of a vegetable crop on the land, which tallies very well with the figures given.

At first sight, it would seem possible to increase the yield per hectare of tambak still further, by administering still more vegetable and chemical manure, but it should be realized that this could too easily lead to oxygen depletion and hydrogen sulphide formation by too intensive biodegradation processes. A higher production could, however, be achieved through a better control of predators, by installing rotating plankton sieves in the sluice gates, and by killing fish and other predators quantitatively with adequate non-persistent pesticides, before stocking the ponds. If a shrimp by-catch is to be maintained, the ponds should be purposely stocked with shrimp fry, in case the natural immigration is blocked with plankton sieves.

The legal structure of the tambak farms shows a typical pattern which many would call truly "capitalistic" or even "feudal": an owner, living far away, not participating in the daily work, but receiving a high share of the net profits; the participant-caretakers, responsible for the daily work, for maintenance and surveillance of their units, receiving one-third of the net profit; and the locally hired casual labour for constructing additional ponds, carrying out larger maintenance works, and for the fishing out of ponds. The latter are part-time workers, partly paid in kind. A closer study of this pattern makes clear that it is, indeed, very well adapted to the local situation: the construction of a tambak requires a noteworthy investment of capital, and even the larger maintenance works and the purchase of fry would go far beyond the financial capacity of the participant-caretakers. It is therefore logical that the owner receives a fair share of the net profit for his capital investment and for the risks involved in purchasing expensive bandeng fry in large quantities. The participant-caretaker, on the other hand, will get a better income if his efforts to control predators and competitors lead to a lower mortality among the juvenile fishes, and a higher yield in his ponds. The casual labourers, whose assistance is invoked only occasionally, are not unemployed in the remainder of the year, but find all kinds of other work to do in these predominantly agricultural areas.

The milkfish farms are based on suitable hydrographic and climatological conditions, on know-how gathered over the ages, on expert advice from the Governmental Fishery Institute, on the availability of labour to be hired on a temporary basis, so that their wages are not weighing full-time on the budget of the enterprise, and on the good market in a densely populated part of the country. A further increase in the production of milkfish is certainly possible, and the market is certainly willing to absorb it for a reasonable price. Further market development, including a high-quality export product, could lead to a further increase in the price of this well-flavoured firm-fleshed fish. A serious bottleneck is, for the time being, a steadily decreasing catch of fry in the areas which always produced the quantities one needed. Air transportation has opened the possibility of collecting fry from far away, but in the future only hatchery-produced fry would eliminate the fear that capital investment for constructing more ponds would, in the long-run, not be remunerative. National or international efforts to solve the intricate problems of maturation and spawning *Chanos chanos* under controlled conditions are highly desirable to aim at an increased milkfish production.

G. BIBLIOGRAPHY

De Jaager, C.J. and Lawick van Pabst, H.J.W., 1903. Rapport nopens de Visserijvorm van Java en Madura. Landsdrukkerij, Batavia, 150 pp.

Delsman, H.C. and Hardenberg, J.D.F., 1934. De Indische Zeevisschen en Zeevisscherij. Visser & Co., Batavia, 388 pp.

Hofstede, A.E., Ardiwinata, R.O. and Botke, F., 1953. Fish culture in Indonesia. Indo-Pac. Fish. Counc. Spec. Publ. No. 2, 129 pp.

Reijntjes, E.J., 1926. De vischteelt in zoutwatervijvers. Mededeling Afdeling Nijverheid, Weltevreden, 69 pp.

Schuster, W.H., 1948. Over tambak garnalen. Landbouw, 20 (11/12): 21—31.

Schuster, W.H., 1949. On the food of the bandeng (*Chanos chanos* Forsk) in Indonesian ponds. Meded. Alg. Proefstn. Landbouw, Buitenzorg, Java, No. 86, 20 pp.

Schuster, W.H., 1952. Fish culture in brackish water ponds of Java. Indo-Pac. Fish. Counc. Spec. Publ., No. 1, 140 pp.

Schuster, W.H., 1952. An annotated bibliography on the culture of milkfish *Chanos chanos* Forskal. Indo-Pac. Fish. Counc., Occas. Pap., 52/3, 24 pp.

Schuster, W.H., 1960. Synopsis of biological data on milkfish *Chanos chanos* Forskal, 1775. FAO Fish. Synop., No. 4, 67 pp.

Sunier, A.L.J., 1922. Contribution to the knowledge of the natural history of the marine fish-ponds of Batavia. Treubia, 2 (2/4): 156—400.

Chapter 4

FARMING PENAEID SHRIMPS IN JAPAN

Example
First stage: Yamaguchi Prefectural Marine Products Seedling Centre at Aeo-Cho, Yoshiki-Gun, Yamaguchi Prefecture.
Second stage: Asahi Shoji K.K. shrimp farm. Kurogata Minami, Aeo-Cho, Yoshiki-Gun, Yamaguchi Prefecture.
Period of observation: 1973.

A. BACKGROUND

A.1. General principles

The Japanese shrimp, locally called "Kuruma-ebi", is reckoned among the highest grade fishery products of Japan. The quantities supplied by the fishery are small, but the demand is much greater, and the price the customer is willing to pay for live shrimps of this species is astonishing. It is therefore easily understood that, long ago, the first attempts were made to farm this top quality crustacean for profit. Shortly after the turn of the century, shrimp farms were created in Kumamoto Prefecture; juvenile specimens, caught in the sea, were stocked in enclosed ponds with sandy bottom and fed until they reached marketable size. This system can hardly be called a semi-culture, rather a form of live-keeping. The production was small and the supply of seedlings irregular.

Now Japanese shrimps are farmed from fertilized eggs, almost in a full-cycle system. At the centre of this development were the tireless efforts of one man, Dr. M. Fujinaga (also spelled Hudinaga), and it is fascinating to read how his first success in laboratory spawning of mature shrimps in 1933 led step by step to the unravelling of the secrets of the biology of this species, and to practical systems of rearing of the larva to a bottom-stage large enough to be stocked in open-air ponds.

The main difficulty was that each developmental stage required its own special food, and successful rearing is only possible when such food is available in reasonable quantities. This usually means that the food organisms must be reared too, especially the small organisms required for the juvenile stages. The technical phase of shrimp farming was reached in 1963 with the setting up of the Kuruma Prawn Farming Co. Ltd. at Ikushima, Takamatsu. High hopes were held of developing a real bio-industry, producing impressive quantities of shrimps on a small acreage, independent of the whims of nature. But soon it was learned that large scale production of various types of food in pure culture would be economically prohibitive, and further, that a benthic organism like the shrimp, virtually lives in a thin layer of water just at the bottom of the ponds and during the daylight hours in the topmost layer of the sediments. All the water of the middle and higher layers remains unused. Unfortunately, it is exactly this bottom layer and the top layer of the sediment itself which are very difficult to keep constantly well oxygenated. It therefore appears impossible to farm the shrimps in dense populations. Even when food can be amply administered, it is impossible to keep more than about 20 fair-sized shrimps per m^2 of pond bottom. Therefore one cannot harvest more than 300—400 g of shrimps per m^2, or 3—4 metric tons per ha. This compares rather poorly with the results obtained in a real bio-industry, and there is little hope for further improvement.

There are now about 14 shrimp farms in Japan, operating with some 100 ha of water sheets, producing in total about 200 metric tons of marketable shrimps per year. Most of the farms are to be found in the warmer southwestern part of Japan, especially on the shore of the Seto Inland Sea, and on the east coast of Kyushu. Most of these farms have been set up where there were derelict salines, and some are constructed in the intertidal zone.

Only a few farmers carry out all the rearing phases themselves; virtually all consider it of great economical advantage to use the Prefectural Seedling Centres, where the juvenile phases are reared on a large scale and where the farmer can purchase juvenile shrimps large enough to be stocked in the open-air ponds.

Even in a well-equipped seedling centre, only part of the eggs spawned by the mature females will hatch, and some losses are unavoidable in the early developmental stages, so that the overall survival, by the time the fry is ready to be transferred to the rearing ponds, seldom exceeds 20% of the original eggs. However, after stocking, the losses are remarkably low in a well-exploited farm, and up to 80% of the number of specimens stocked may be harvested after 5 or 6 months of farming. Since the shrimps stop growing when the water temperature drops under 15°C, the whole batch should be sold before the end of the year. To obtain marketable shrimps of a good size, the ponds should be stocked in May or June, which means that the seedling centres should utilize the onset of the natural spawning season, which starts in April and lasts until September. Only one crop per year is produced and for several months the ponds are out of use. The farmed shrimps should arrive alive on the market, for only live shrimps fetch a very high price. This requires special precautions. Farmed shrimps make a price 20% higher on the market than wild shrimps caught by the free fishery; the former are less damaged and are better in flavour than the wild specimens. This latter aspect should be ascribed to the fact that the farmers administer high quality food to their shrimps, predominantly consisting of live little-neck clams, partly also of Mysid shrimps. On the other hand, this is the main reason for concern in shrimp farming, for little-neck clams are not only a valuable item as human food, but their supplies are also declining in many places, due to overfishing and pollution.

A.2. Biology of the species

The Kuruma shrimp, *Penaeus japonicus*, belongs to the Penaeids, a group of swimming decapod crustaceans characterized by an elongate laterally compressed abdomen carrying a fan-like tail, by the rostrum on the head, and by pleopodes serving as swimmerets — except the first pair, which in the male has developed into a copulation organ. The Penaeids can swim, but live close to the bottom. During the daylight hours they bury themselves in the sand to avoid their predators. They are therefore obligatory nocturnal feeders. They live on small bottom invertebrates, occasionally on carrion and detritus, but cannot overpower larger bivalves, duly protected by their shells. Experience in the shrimp farms has demonstrated that penaeid shrimps are not very efficient as food convertors, for it takes 10—15 kg of food to produce 1 kg of shrimp. It is surmised that this poor efficiency is caused by the considerable loss of energy in repeated moulting. Food passage through the

Fig. 39. The Kuruma shrimp, *Penaeus japonicus*, of marketable size, brought ashore in a keep-net. (Photo P. Businger.)

intestinal tract is rapid, but especially in denser populations, some recycling takes place due to the coprophagous tendencies in Penaeid shrimps.

A female Kuruma shrimp of 15 to 18 months old weighs 50—120 g. Immediately after moulting, a male specimen copulates with her and brings its sperm, contained in a spermophore, into her receptaculum, so that she can make use of live sperm at any time, until she moults again. Copulation may take place at any time of the year, but the spawning season lasts in Japanese water from April till September. In the middle of the night the gravid female with violent movements releases her fertilized eggs; a total 500 000 in a fair-sized specimen. The embryos in the fertilized eggs develop rapidly, and some 13 to 14 h after spawning they hatch, releasing a minute *Nauplius* larva, displaying a spasmodic swimming behaviour. The nauplii are still provided with sufficient yolk to continue their further development without feeding. At water temperatures over 25° C they will moult six times within 36 h, becoming a typical *Zoea* larva after the last moult. This is the most delicate phase of larval development, and when the *Zoea* larvae do not find sufficient food they will perish. They swim quickly in a forward direction and feed by preference on small phytoplankton organisms. Minute diatoms such as *Skeletonema costatum* and flagellates such as *Isochrysis galbana* and *Monochrysis lutheri* are excellent food for the *Zoea* larva, but *Chlorella*, though ingested, cannot be digested by them. In about 4 days the *Zoea* moults 3 times to become a larva of the mysid type, somewhat shrimp-like in appearance, standing on their head while swimming, and making backward darts occasionally. They gradually switch over to other food, becoming carnivorous, and should find small planktonic crustaceans to feed on. The mysid phase lasts about 4 days, in which time three moults take place, and finally the juvenile shrimps reach the post-larval phase, now closely resembling their parents, and living like them, on the bottom. They remain carnivorous but can gradually feed on larger food items. If they do not find enough food they will switch over to cannibalism which may lead to serious losses.

They continue moulting, though with decreasing frequency, in all 20 to 22 times. Initially they feed day and night, but from a size of 20 to 30 mm onwards, they feed during the dark only, hiding themselves in the sand during the dangerous sunlight hours. Post-larva of 20 to 25 days old measure 15 mm in length and weigh only 20 mg, but some 5 months later they may have reached the marketable size and a weight of 15—25 g.

The Kuruma shrimp is rather euryhalinic and can be grown in water of 27 to 35‰. Its temperature range is 15 to 33° C, optimal from 28 to 30° C. It can survive at under 15° C but stops feeding, growth and moulting, and at 12° C even swimming and jumping.

A.3. Geographical position

The Yamaguchi Prefectural Marine Products Seedling Centre and the Asahi Shoji Shrimp Farm are located close together near Aeo-Cho in

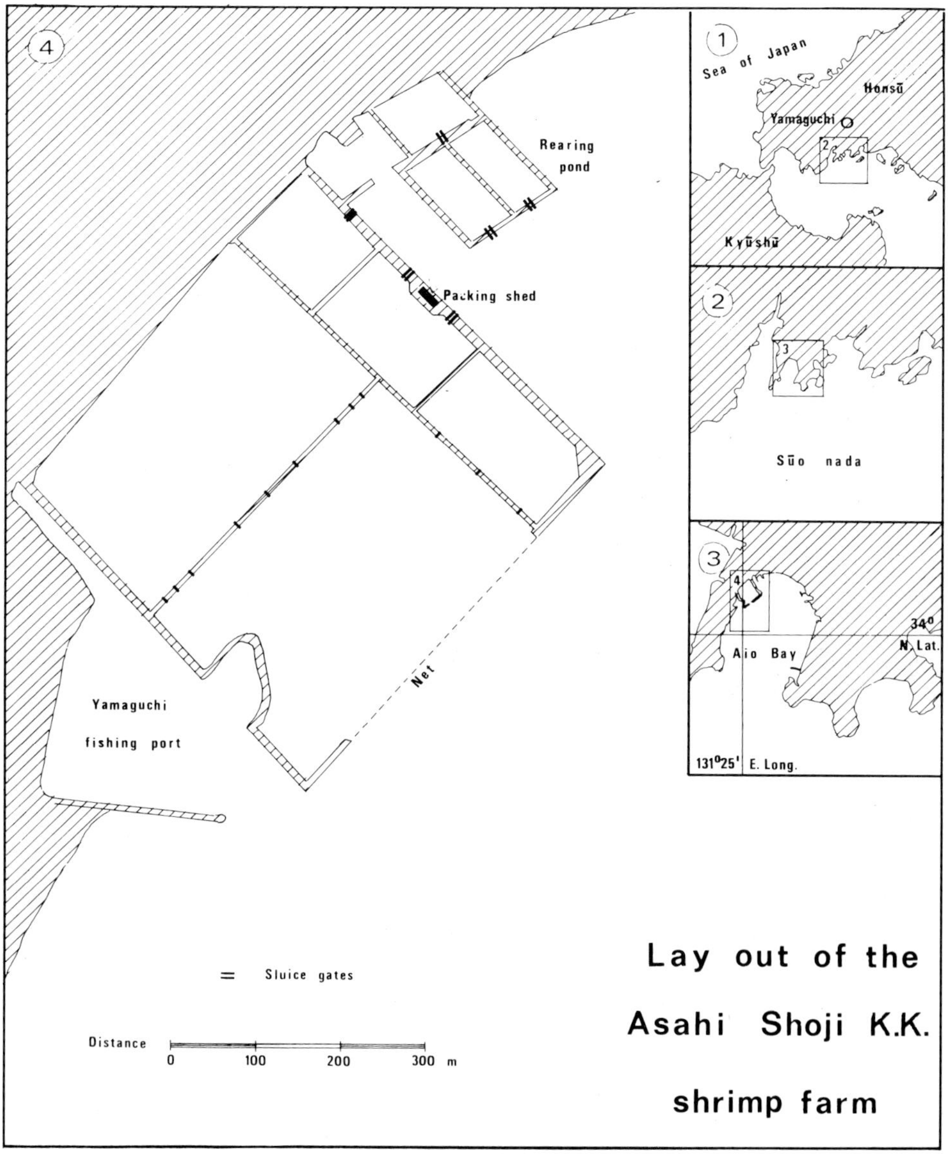

Lay out of the Asahi Shoji K.K. shrimp farm

Yamaguchi Prefecture, on the southwestern tip of Japan's main island Honshu, not far from the city of Yamaguchi (see chart here and chart in Volume I, Chapter 1). The Seedling Centre is on the western shore of an irregularly shaped peninsula, facing the bay called Aeo-Ko, which is continuous with Suo Nada, the western prolongation of the Seto Inland Sea. Its exact geographical position is 34°00′05″N and 131°26′10″E. The Asahi Shoji shrimp farm is located on the northwestern coast of the same bay, at the exact geographical position 34°00′40″N and 131°25′00″E.

A.4. Hydrographic pattern

I. Indoors: the hatchery tanks. In the early days of shrimp farming, gravid females were placed in series of tile-lined indoor tanks which were filled with filtered sea water which was duly aerated and temperature-controlled. The developing larvae received, at the appropriate time, a dose of a mono-culture of phytoplankton, specially reared for that purpose. Though this system is still widely used, a switch is being made to larger units, requiring less labour. The Yamaguchi Prefectural Seedling Centre grows its shrimp larvae in large concrete tanks (three tanks of 8 × 8 × 2.5 m in the one rearing house, and five of 7 × 7 × 2 m in the other). The concrete walls of the tanks are coated with plastic paint. A superstructure of steel and glass has been constructed over these sets of ponds.

The bay facing the Seedling Centre contains clear sea water of high salinity over a clean, sandy bottom. The sea water is pumped electrically to a filter house through vinyl pipelines 20 m long, constructed in pairs for alter-

Fig. 40. The large indoor tanks for rearing shrimp fry in the Yamaguchi Prefectural Seedling Centre. (Photo P. Businger.)

nate use. In the concrete filter house the water has to pass through a column of sand before it is led into the rearing basins. This filters the water adequately, and prevents the introduction of predatory organisms. The sand in the filter has to be replaced once a year.

An oil-fired central heating system brings the water to the desired temperature of 26 to 28° C; steam passes through heating pipes installed close to the bottom of the tanks, but at some distance from the side walls. A compressor connected to a system of plastic tubes leading to porous air-stones (about one for every 3 m^2 of bottom surface) is used to aerate the water in the tanks. To provide the required food for the larva, the water is enriched with a daily dose of nutrients in an effort to grow a mixed crop of phytoplankton, a natural assortment which should offer the shrimp larvae sufficient food to ingest and digest.

II. Outdoors: rearing basins. In the early days of shrimp farming in Japan, rather small concrete open-air tanks, provided with sand bottom, were generally used for rearing young shrimps to marketable size. Though such small tanks do offer certain advantages, a switch is being made to larger units requiring less labour. Most of the rearing basins now used in shrimp farming in Japan were originally constructed as salines. They vary from 1 to 10 ha, 3 ha being considered as optimal. A shrimp farmer should have at least two outdoor basins at his disposal.

The outdoor basins are cut off from the sea, but should be regularly flushed: one-fifth to one-quarter of the water should be replaced daily. The tides, which range here from 2.5 m at spring tide to 0.80 m at neaps, are used for this. The water is let in through a sluice gate, or through a siphon, but before entering the rearing pond it comes into a special section surrounded by finely meshed nylon netting screens of solid structure. This is necessary to prevent shrimps reared in the pond finding their way to the sea, and also to prevent predatory organisms entering the rearing ponds. The ponds carry 1.5 to 2 m of water at high tide, but 1.0 to 1.5 m at low tide. Subsectioned ponds should be provided with screens of nylon netting reaching about 60 cm above the water level, to prevent shrimps from jumping into another section.

The hydrographic conditions of the water in the outdoor basins is a matter of considerable concern. Since the shrimps hide in the sandy bottom during daylight, special attention should be paid to the conditions in the topmost sediment layer, and in the water immediately above the bottom. Since flushing by the tides does not lead to strong currents, the water in the basins can become stratified, especially when there is no strong wind blowing. The sun may heat the surface water layer more than the deeper ones, and rainfall may lead to a lower specific gravity in the top layer, which also promotes stratification. Therefore, the deeper water layer often shows a different hydrographic pattern to the higher layer. The situation is aggravated because, not only do the shrimps themselves consume oxygen from the deeper layer, but also large quantities of food have to be administered, part of which gets dissolved in the water. Microbial degradation leads both to serious oxygen consumption and to a dense phytoplankton, making use of the released nutrients. Especially towards dawn, the situation may become critical for the shrimps. Sometimes the colour of the water changes abruptly from coffee-brown to whitish, an indication of a fatal development caused by oxygen depletion, hand in hand with development of excessive quantities of the poisonous components ammonia, hydrogen sulphide and carbon dioxide.

In fact it is the quantity of oxygen which forms the bottle-neck, and determines how dense a stock of shrimps can be grown in a rearing basin.

There is not much one can do to improve the situation. Water temperature, so important for active feeding and growth of the shrimps in the open air basins, depends completely on insolation. In January the temperature of the sea water will be about 10° C, February 7° C, March 13° C, early in April the biological critical level of 15° C will be surpassed, in May the average temperature will be about 16.5° C, June 18.5° C, July 24° C, August 27° C, September 26° C, October 22.5° C, November 17° C, dropping again under the 15° C level late in November or early December. These figures hold good for the sea water in the bay, but in the shallow basins one may expect a more rapid rise in spring and also an earlier decline later in the year. These figures indicate that one can expect growth of the shrimps from April to December, and not in the colder winter months.

The salinity may become reduced by heavy rainfall. This can lead to stratification when there is little wind, but the reduced salinity as such, does not seem to do much harm to the shrimps, living in deeper layers.

Oxygen content is a matter of serious concern. A dense vegetation of phytoplankton will see to oxygen production during the daylight hours, but contributes to oxygen consumption at night. Food-remains in liquid or particulate form require large quantities of oxygen for their microbial breakdown. A study of oxygen consumption in a shrimp pond led to the conclusion that some 8% of the oxygen consumed per day should be attributed to the shrimps themselves, about 7% to fish living in the ponds, 15% to the processes taking place in the sandy bottom, whereas up to 70% of the consumption should be ascribed to the activities of microorganisms, phytoplankton included, dispersed in the water. Oxygen values under 3.5 p.p.m. should be considered as dangerous for the shrimps. Since the shrimps wish to hide in the sand during daylight, special attention should be paid to the conditions prevailing there. Even when the bottom consists of the preferred rather coarse sand it may be that there is a black sediment, rich in hydrogen sulphide, at a depth of only a few millimetres, which makes it difficult for the shrimps to dig in. Food remains and microbial activities easily lead to such a situation.

In the smaller tanks, formerly used on large scale, a false bottom of finely meshed nylon netting was often installed, on which one deposited a layer of 2 cm of coarse sand. A series of stand-pipes was installed on the true bottom, popping up above the sand of the false bottom. An air lift forced water from below the false bottom upwards through these stand-pipes and in its turn this water had to find its way down again through the sand layer in which the shrimps live during the day. This percolation system leads to better conditions in the sand than would be observed in tanks without a false bottom. However well this system may have worked, it has now generally been abandoned, to switch over to larger basins, these being easier to fish out with nets and less costly in construction and maintenance. To promote aeration, and to prevent stratification, paddle-wheel aerators are placed in the basins. The low rate for electricity at night makes it possible to run such devices continuously at night; when necessary they are also switched on during daylight. The shrimp farm under consideration does not use normal paddle-wheels, described in the chapter dealing with eel farming in Japan, but rather a contraption knocked together from the rear axle of a car and a large 6-blade propeller, all this installed on drums, used as floats, and driven by an electric motor. But even then the balance between oxygen content, stock density and food administration should be carefully controlled. In the Asahi Shoji farm one 5 h.p. aerator is used for every 5 000—7 000 m^2 of pond.

A.5. Legal aspects

The Yamaguchi Prefectural Marine Products Seedling Centre is a Governmental establishment. Their aim is consistent results and efficient production

Fig. 41. Pond aerator on drum floats as used in the Asahi Shoji shrimp farm.

of shrimp seedlings, and they sell those to the farmers at a fixed rate. The running costs for producing seedlings may be covered by the price, but amortization of buildings and installations is for the account of the Government, which also has provided the funds for research in the initial phase of shrimp farming. There are also co-operative seedling centres, e.g. at Aeo, and there are still a few shrimp farmers who produce their own seedlings, but the majority of farmers have come to the conclusion that it is economically more attractive to buy them from a Seedling Centre.

The shrimp farm under consideration is a private enterprise named Asahi Shoji K.K. The acronym "K.K." stands for Kabushiki Kaisha", which means that it concerns a company with limited liability. The grounds on which the company operates are private property. There is occasional police surveillance, but it is primarily the company guards who watch over the property.

B. SEQUENCE OF OPERATIONS

B.1. Acquiring parent stock

It has been demonstrated that it is possible to overwinter specimens of the Japanese shrimp grown from the egg, and to rear them until they become mature. Scientists in other countries, working with other species of penaeid shrimps often experience difficulties to reach this goal. Since shrimp rearing in open-air ponds is largely dependent on the natural water temperatures, the operations have to be organized in such a way that the shrimps reach the marketable size before the end of the year. This means that one should concentrate on the first part of the natural spawning season, which locally lasts from March to September. In captivity, maturation is not yet reached in the spring months and therefore one is still completely dependent on mature females caught in nature as parent stock.

Japanese fishermen operating with trawls in the Suo Nada, the western part of the Seto Inland Sea, catch full-grown shrimps at depths from 10 to 40 m. They are found willing to select mature females from among their catch, and to take the trouble to carry those alive to the Seedling Centre. From March to September they may find females carrying ripe eggs under their dorsal carapace; up to 10% of the catch. An experienced eye can see, against the light, whether the eggs are really close to spawning. Such gravid females, having been fertilized just after their last moult, weigh from 50 to 120 g. The size of females carrying eggs is found to be declining with the advance of the season, during which the fishing gradually shifts from the open Pacific towards the Inland Sea.

The live shrimps can best be transported in the fish-well of the ship or in a tank with aerated sea water. It is even possible to transport live shrimps out of water, packed in chilled cedar saw-dust, but this latter system is said to yield poorer results.

The Yamaguchi Prefectural Seedling Centre buys, in the course of the season, 100 kg of gravid females from the fishermen. In other countries fishermen are less co-operative, perhaps because they consider that farmed shrimps may compete with the free catches, the more since the farmer makes a better price than the fisher gets for trawled shrimps. The Japanese fishermen may hold a different view, since the various Prefectural Seedling Centres not only rear various species for farming purposes, but also for stocking the Inland Sea, from which the yields of valuable species has declined, though its general productivity remains unshaken.

B.2. Spawning in the indoor tanks

In the Seedling Centre, the indoor tanks are filled with filtered sea water of 32 to 35‰ salinity which is heated to 26—28° C prior to stocking them with gravid females. The sea water is dosed daily from the beginning with nutrients: 2 p.p.m. nitrate-N, plus 0.1 p.p.m. phosphate-P, some potassium silicate is also occasionally administered to the water in the tanks. The purpose of this enrichment is to encourage the development of a mixed crop of small diatoms and flagellates. The glass roof transmits enough sunlight to promote the growth of diatoms, and in due course the water acquires a brownish colour.

The phytoplankton is to serve as food for shrimp larvae in certain developmental stages and provides, at the same time, the shade the larvae seem to prefer. It is hoped that the phytoplankton in the tanks will be qualitatively of a good composition, but there is little one can do if too large or undigestible phytoplankton organisms are found to predominate in the natural assortment. Evidently the efforts usually meet with success, so that the old elaborate system of growing phytoplankton organisms in pure culture and adding a daily dose to the rearing tanks can be avoided. Seeding the tanks from the beginning with a culture of the minute diatom *Skeletonema costatum* is said to be an appropriate measure.

It is necessary to keep the tanks as clean as possible and to avoid all contact with potentially dangerous material. Therefore the bottoms of the tanks have been constructed in such a way that they gently slope towards one corner, so that the tank can be emptied completely. The use of metal parts is avoided; all tubes, valves and drain pipes are made of non-toxic plastic. Compressed air, introduced via plastic tubing and "air-stones", sees to a mild water circulation and to good aeration. In selecting a compressor, a type such as the "Root-Blower", which gives air practically uncontaminated with lubricating oil, is preferred.

The Yamaguchi Prefectural Seedling Centre has in one of its breeding houses three basins of 150 m^3, in the other five basins of 100 m^3.

The gravid females are put in the tanks, preferably in the late afternoon, some 50 to 60 together in a tank of 100 m^3. Spawning takes place in the middle of the night: the female releases her eggs with some vigorous spasmodic movements. However good the selection of gravid females may be, one should not expect that all of them spawn the first night. Dead or spawned out females should be removed by scoop-nets the next morning, but the others should get another chance the next night. After that, all are taken away, since experience has shown that no further spawning will take place thereafter. In practice about 50% of the introduced females will actively spawn. One female may release from 300 000 to 1 000 000 eggs, and usually some scum appears at the surface of the tanks indicating that spawning has actually taken place.

B.3. The Nauplius phase

The *Nauplius* larvae hatch 14—15 h after spawning. Not every egg yields a nauplius, a good average being about 50%. The nauplii are provided with yolk and need not take any food in going through their regular series of moults, finally leading to the *Zoea* stage. Nevertheless, the daily dose of nutrients continues to be administered, but after about the 10th day it can be judged from the colour of the water, held against the light, whether the dose can be halved.

B.4. The Zoea phase

Originally there were 50 000 to 100 000 eggs per m^3, half of which yielded nauplii. One should be content if 36 h later about half of the nauplii transform to *Zoea*, after their 6th moulting. The zoea phase is said to be the most delicate. The zoeas should find enough readily digestible food, and since they cannot hunt for it, the density should be quite high. Flagellates such as *Isochrysis galbana* and *Monochrysis lutheri* are excellent as food for zoeas, but also the minute diatom *Skeletonema costatum* is good food. An organism like *Chlorella*, on the other hand, is undigestible. One thousand *Skeletonema's* per millilitre is an appropriate density. It is possible to rear *Skeletonema* in pure culture and to concentrate it for storage, but the technique now used in the Yamaguchi Prefectural Seedling Centre is to produce a mixed crop of phytoplankton in the tanks, and at most, to seed these tanks with a *Skeletonema* culture to promote its predomination.

B.5. The mysid stage.

Part of the zoeal survivors will be aborted during mysid moults. The mysid larvae look like minute shrimps. They feed initially on the same type of phytoplankton as the zoeas, but gradually switch over to carnivorous habits.

When small tanks were used in the early phases of shrimp farming, the most suitable food for the mysid larvae was nauplii of the brine shrimp, *Artemia salina*. It was, in fact, during Dr. Fujinaga's time in the United States that he became familiar with *Artemia* eggs, which can be stored dry for a long time to produce nauplii by immersion in sea water at any moment one needs them, and this led to a breakthrough in experimental shrimp farming. *Artemia* eggs can be purchased in the United States, but the quality may vary greatly. At the Yamaguchi Centre, *Artemia* eggs from Utah are used, the nauplii yield being 80%. Five grams of Artemia eggs per day are needed per 10 000 shrimp larvae, or 500 g per day per tank of 100 m^3. The dry *Artemia* eggs are simply scattered in the rearing tank. Since *Artemia* eggs are quite expensive, substitute foods were sought. The small rotifer *Brachionus plicatilis* appeared to give good results. One can grow *Brachionus* rather easily on a diet of yeast cells, or on the minute alga *Dunaliella*. In the Yamaguchi Seedling Centre, a culture of 50 *Brachionus* per ml, if properly fed with yeast, shows a density of 300 per ml shortly after. These can be filtered off with a plankton net, to be administered to the rearing tank in quantities adapted to the number of mysid larvae. Special tanks are not required to grow *Brachionus*; it suffices to partition off a section of the big rearing basin. As a rule a spontaneous growth of various small zooplanktic crustaceans is also observed in the tanks. The water with which the tanks were originally filled evidently held enough seeds of various kinds to make a zooplankton bloom possible in the coffee-brown diatom soup. Extra water is often added to the tanks, now filtered through a plankton net instead of through a sand column, to promote the development of small zooplankton organisms. If the mysids still appear to run short of food, finely minced clam meat or chopped annelids can be fed to them. There are 20 000 to 30 000 shrimp larvae per m^3.

B.6. The post-larval stage

After about four days of mysid life, the post-larvae appear, real little bottom-stage shrimps, with truly carnivorous and coprophagous habits. The water temperature can now be lowered to 22—25°C without inconvenience, but attention should be concentrated on correct feeding. The post-larvae are voracious, and when they do not find enough food they do not hesitate to eat their kin, which may lead to serious losses. *Artemia* and/or *Brachionus* can be used as feed for some time, but some other food must be used in addition: minced bivalve meat (asari, i.e. *Venerupis philippinarum*) or minced annelids; some 10% of the body weight of the juvenile shrimps per day. This comes to 20 g minced food per day for 10 000 juvenile shrimps, to be increased to 80 or 100 g after 3 or 4 more moults, on about the 20th day. Though feeding should be generous to avoid cannibalism, overfeeding easily leads to fouling, and hence to low oxygen values and losses among the juvenile shrimps. At this stage the number of juveniles left over will be about 20% of the number of original eggs.

The post-larvae are kept in the indoor rearing tanks till they reach a length of 12 to 13 mm and a weight of 0.01 to 0.02 g. Under good rearing conditions 5 000 to 8 000 will be found per m^3. Then, after 25 to 30 days of post-larval life, they are ready to be transferred to the outdoor basins.

B.7. Collecting the seedlings from the indoor tanks and transfer to the rearing ponds

When the seedlings are ready for transfer to the outdoor ponds they have to be collected from the indoor basin. They are not fished out with nets; another system is used which leads to minimal damage. The indoor basins are provided with drain-pipes which lead to the outside of the building, where plastic valves are installed to prevent loss of water during the indoor rearing phase. The water can be drained from the basins by gravity by opening the valves. First a screen of finely meshed nylon gauze held in a frame is placed around the inside aperture of the drain-pipe, next the outside valves are opened, so that part of the water can flow away without taking any juvenile shrimps. Next a rectangular box of nylon netting, (Fig. 42) held in a wooden frame, is placed under the outlet, the inward screen is taken away and the drain-water is allowed to pass through this box, which holds back the seedlings. The juveniles are kept immersed during this whole procedure. From time to time the contents of the gauze box are emptied into a container.

For transfer to outdoor rearing ponds in the immediate vicinity, a 0.5 m^3 transportation tank, filled with water, can be used. This is placed on a fork-lift truck. Transfer over greater distances requires special care, for under crowded conditions, cannibalism leads to drastic losses. To avoid this, low temperature water is used to reduce the activity of the shrimps. Five thousand to 10 000 seedlings are placed in 20 l plastic bags containing 6—8 l of sea water, and oxygen is added. The bags are placed in cardboard boxes, and in this manner a 2-ton refrigerated truck can safely take 600 000 to 700 000 seedlings for a trip of up to 12 h. One should gradually raise the temperature of the water again, to prevent a thermal shock when releasing the seedlings in the rearing ponds.

B.8. Preparing and stocking the rearing ponds

The Asahi Shoji farm operates with eight ponds of various size, the smallest measuring only 1 500 m^2, the largest as much as 77 000 m^2. There is a separate group of three ponds averaging 6 000 m^2 each, and next to the large pond is a row of three others measuring 11 000, 16 000 and 18 000 m^2. The latter ponds are constructed against a long pier, which shows centrally, a widened section carrying the packing shed. To the seaward side of the great

Fig. 42. Wooden frame with nylon netting for collecting juvenile shrimps from the hatchery tanks. (Photo P. Businger.)

pond complex one finds a semi-enclosed area, which is screened off from the sea with double netting, one of steel, the other — which is exchangeable — of nylon. A breakwater outside the netting provides the necessary shelter.

The bottom of all the ponds is sand. The partitions between the ponds consist of nylon netting fences, high enough to prevent the shrimps from jumping over (Fig. 44). The ponds are flushed by the tide. Sluice gates giving direct access to the sea are duly screened off at the entrance to avoid escape of shrimps and entrance of predators.

It is advisable to prepare the pond bottom carefully before stocking the seedlings. The pond should be drained as well as possible, and derris or rotenone used to kill any fish left behind in deeper pools. Next, the sandy bottom should be ploughed and levelled, bringing in more sand when necessary. Thereafter, the ponds are slowly filled again, first of all the nursery pond, one of the smaller sections in which the seedlings are to be stocked. Usually up to 150—180 seedlings, weighing 10 to 20 mg each are introduced per m^2 of nursery pond, and their even spreading over the whole surface of the pond has to be controlled. Stocking is performed from early in May to mid-June.

Fig. 43. The packing shed of the Asahi Shoji shrimp farm, built on a widened section of the pier. In the foreground one of the sluice gates. (Photo P. Businger.)

B.9. Feeding the shrimps in the outdoor ponds

Though the shrimps do find some natural food in the ponds, they should be fed liberally to make them grow fast enough to be marketed at the end of the year. The Japanese shrimp farmers use exclusively natural food items, though the coprophagous tendencies of the shrimps indicate that they must be willing to accept formula food.

Practical experience has shown that a mixture of fresh bivalve meat and deep-frozen soft-shelled crustaceans lead to the best results. The bivalve commonly used to feed the shrimps is the little-neck clam *Venerupis philippinarum*, locally called "asari". At times of scarcity, one may revert to meat of mussels (*Mytilus*) or oysters (*Crassostrea*). The shrimp-like crustaceans fed belong predominantly to the genera *Neomysis* ("ami"), *Plesionika* ("Tokushima ebi") and *Metapenaeopsis* ("aka ebi"). The clams are purchased from local fishermen, who collect the clams with a toothed dredge drawn behind a boat. Rich asari beds are found in the coastal waters of the Isle of Kyushu. Overfishing and pollution have made the asari decline or disappear in many a site. The crustaceans are purchased as deep-frozen blocks from places such as

Fig. 44. The partitions between the ponds of Asahi Shoji shrimp farm consist of nylon netting fences, high enough to prevent the shrimps from jumping over. (Photo P. Businger.)

Shiogama and Miyagi Prefecture. They may come from the North Pacific. The asari is also used as human food, especially the larger specimens, and are hence quite highly priced. The small shrimps administered as food are, nowadays, not used as human food on an appreciable scale. In times of scarcity, squid or scrap fish may also be mixed into the diet for the shrimps reared in the ponds.

The shrimps cannot overpower live bivalves themselves, so the bivalve flesh must be made accessible to them. A bivalve crusher (Fig. 45) has been developed. Basically it consists of two spring-mounted steel rollers, each about 1 foot in diameter, placed together with their axes in the same horizontal plane. With the inward rotation of both rollers the clams are crushed by the tension on the roller springs. One can feed the crushed clams as such, and leave it to the shrimps to pick out the edible parts, but when large amounts of crushed shells are administered, small pieces of broken shell may in the long run harden the texture of the bottom so much that the shrimps find it difficult to dig in during daylight. Therefore, often the crushed clams are passed through a meat separator, a metal container with rotating arms

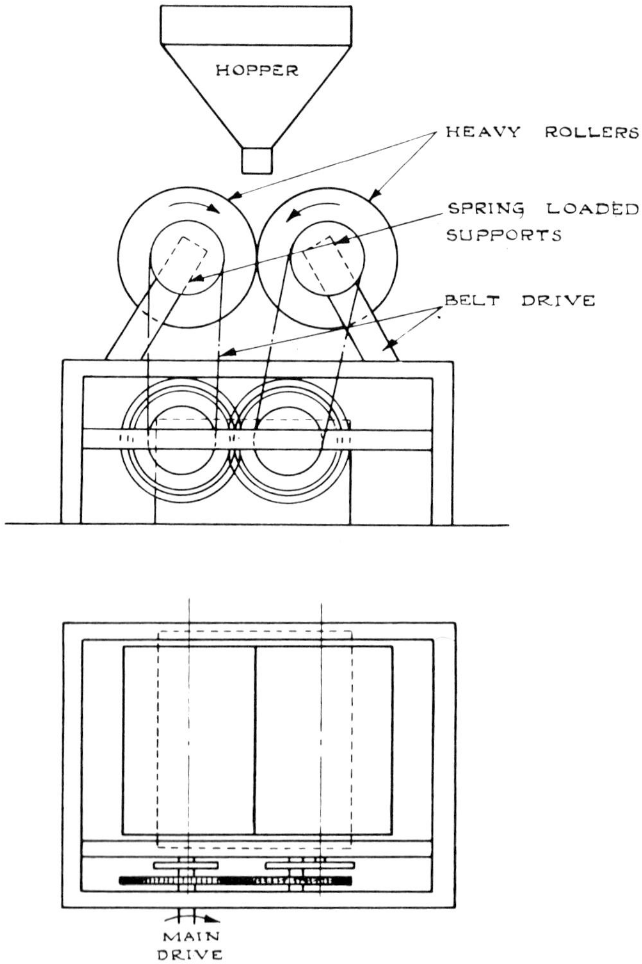

Fig. 45. Clam crusher in side-view and top-view.

moving at about 20 r.p.m., to stir up a mixture of water, clam flesh and shell debris. A steady flow of excess water carries away the clam flesh, which is then collected in a sieve arrangement. But it seems that part of the nutritious material gets lost this way and more clams are needed to obtain a given weight increment of the shrimps than when the crushed clams are administered as such. *

The food administered in the nursery pond, where the smallest shrimps

* It is interesting to note that practically the same apparatus was developed in The Netherlands during the war to crush the clusters of *Crepidula fornicata*, and to separate the meat from the broken shells.

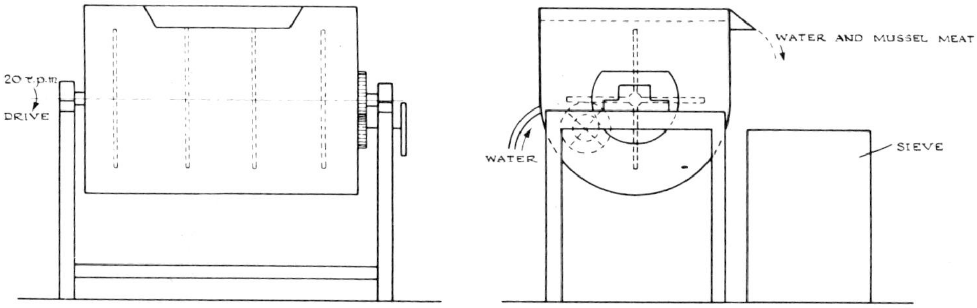

Fig. 46. Apparatus to separate the mussel meat from the shell debris in front-view and side-view.

are reared, should always be finely chopped. In this phase, the food is given 2 to 5 times per day. From the moment the shrimps have reached an individual weight of 1 g, the food is administered in one dose per day. It is given in the late afternoon, and usually consists of crushed bivalves.

Contrary to expectations, fresh bivalve meat appeared to be a more complete food than minced deep-frozen crustaceans. Since little-neck clams are quite expensive and often difficult to get in the quantities required, as a rule 30% bivalve meat and 70% crustaceans are used as food. The smallest categories of shrimps are voracious and wasteful feeders, and require quite a high dose of food; in the first few days up to 200% of their body weight per day, shortly after 50%. When they weigh 1 g they should receive 25% of their body weight in feed per day, to be lowered to about 10% when they have reached 10 g. At water temperatures of 12° C and lower, the shrimps do not take any food. Food conversion is not very efficient in shrimps: on an average one should count on 13.5 kg of feed, shells not included, for the production of 1 kg of shrimp.

The Asahi-Shoji shrimp farm needs a daily supply of about 6 metric tons of live clams, rising to a maximum of 30 metric tons in the height of the season, and 2 metric tons of deep-frozen crustaceans. The actual composition of the food administered depends largely on the supply. When a high percentage of crushed clams has to be offered, it becomes necessary to remove shell remains from the bottom of the ponds from time to time.

Administering the food to the larger shrimps in the ponds takes one man about 30 min/ha each day. Preparation of the food takes more time.

B.10. Growth, transfer to larger ponds, and survival

The shrimps in the nursery ponds grow quite fast in summer, and require thinning out after about one month, for there should never be more than about 250 g of shrimps per m^2. The best system for catching the young shrimps for transfer to the larger rearing ponds, is by making use of pound nets. Such pound nets are set up, being duly anchored, in the nursery ponds at some distance from the shore. A wing net reaching out from the shore guides the shrimps, moving around during the dark, to the body of the pound net. They may end their migration in the cone-shaped traps, from which they are taken up in the morning to be transferred to the rearing ponds. There they are stocked at a density of 15—20, or at very most 25 individuals per m^2. To make sure that the ponds are not overstocked, which would lead to oxygen deficiency, it is necessary to count the shrimps on transfer.

Growth is fastest during the summer months, July—September, and the earlier the shrimps are transferred to the rearing ponds, the higher the weight

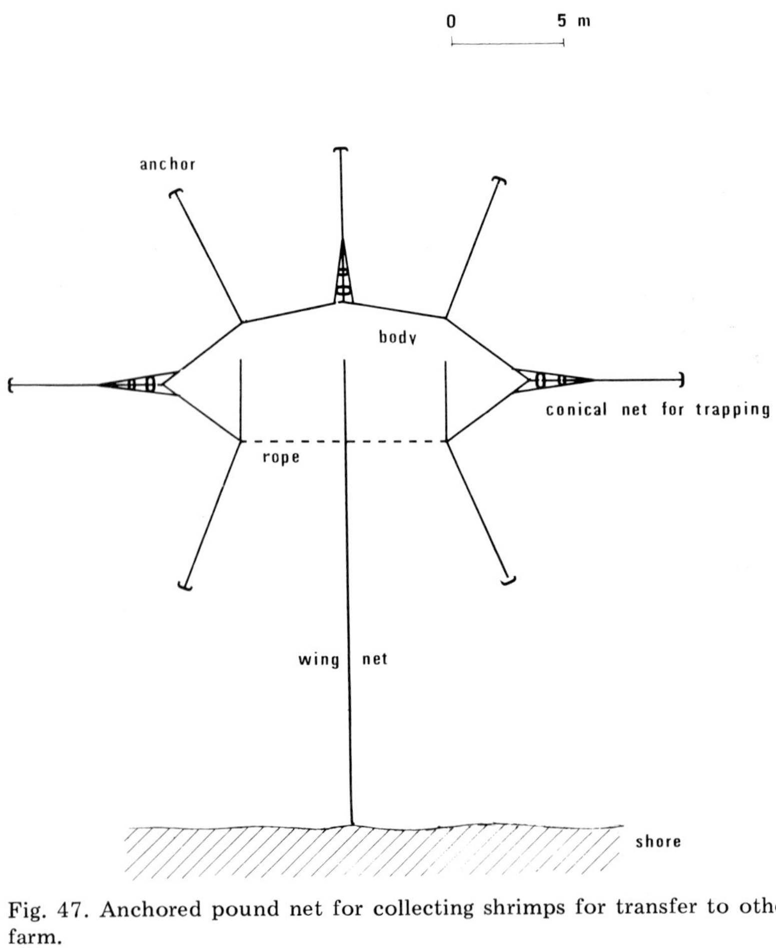

Fig. 47. Anchored pound net for collecting shrimps for transfer to other sections of the farm.

gained towards the end of the year. If 1—2 g shrimps are transferred early in July they will easily reach a weight of 25 g late in October, but if the transfer of 1 g of shrimps is delayed till the middle of August, they may only weigh 12—15 g at the end of the year. The market prefers shrimps of 20—25 g each.

The survival rate may be quite high in the rearing ponds. From the shrimps transferred at an individual weight of 1 g, 80% are harvested at the end of the year, sometimes even more, provided they are properly fed and duly protected against predators. If survival rates are calculated from the moment the seedlings, weighing 10—20 mg each, are stocked in the nursery ponds, the figures fluctuate more widely, but a survival rate to marketing size of 60% is by no means exceptional.

B.11. Harvesting the marketable shrimps

Early in the season marketable shrimps may be harvested with pound nets as have been used in the nursery ponds. This system has the advantage that the shrimps caught show a minimum of damage. As water temperature drops in autumn the shrimps reduce their nocturnal migrations. At lower temperatures they are still quite immobile, dug into the sandy bottom. Therefore moving gear is to be preferred at the approach of winter.

In the Asahi Shoji shrimp farm the shrimps from the rearing ponds are collected with a net mounted on a metal frame equipped with a pipe with nozzles, through which water can be squirted at high pressure to chase the shrimps out of the sandy bottom. They jump up and can be caught in the conical net. This net is towed behind two rectangular plastic catamaran-rigged boats (measuring about 3 × 1 m) propelled with an outboard engine, at a speed of 20—30 m per minute. The frame of the net, made of 12 mm iron bars, is equipped over its whole breadth with a 60 mm diameter water pipe bearing a series of nozzles. Some 0.3 m^3/min of water can be squirted out under a pressure of 4 kg/cm^2, produced by a pump with petrol engine installed in one of the boats. The jets of water stir the bottom to a depth of 6—8 cm and frighten the shrimps out of it. They jump up and are caught in the net moving forward. They are taken out by hand to be put in simple keep-nets such as used by sport-fishermen, hung overboard from the boat.

Some shrimp farmers use electric pulses instead of water jets to chase the shrimps out of the sand. They have installed an electric cable in front of the leadline of the net. Such a net is often drawn by a man wading in front of it, after the pond is drained to a low water level.

B.12. Shipping live shrimps

Immediately upon arrival at the headquarters of the company, the shrimps are weighed and put, still in the keep-nets, in a tank with sea water up to 8° C

Fig. 48. Two plastic boats, rigged as catamarans, fishing for marketable shrimps. The pump thrusts water down the hose (right) leading to the nozzles of the waterpipe of the net. At left, in the keep-nets, shrimps which have been caught. (Photo P. Businger.)

lower than that prevailing in the pond. This is to prevent moulting and to acclimatize the shrimps somewhat before lowering the water temperature still further, to 12° C. This is only necessary in summer, for in winter the temperature of the sea water will be low enough by itself. The shrimps are kept in the tank till the next morning and are then transferred to a packing room, in which the air temperature is kept at 5 to 10° C. They are carefully put in cardboard boxes containing 2.5 kg (summer) or 5 kg (winter), layer by layer, each layer duly covered by chilled sawdust of the Japanese cedar (*Cryptomeria japonica*), locally called "sugi". Extra ice is administered in summer. This sawdust has excellent insulating properties and shrimps, thus packed, will stay alive for 2 days in summer and 4 days in winter. Only live shrimps will bring a top price on the market.

A truck brings the boxes with live shrimps to the nearest airport (Fukuoka), from which they are transported to big cities such as Tokyo and Osaka.

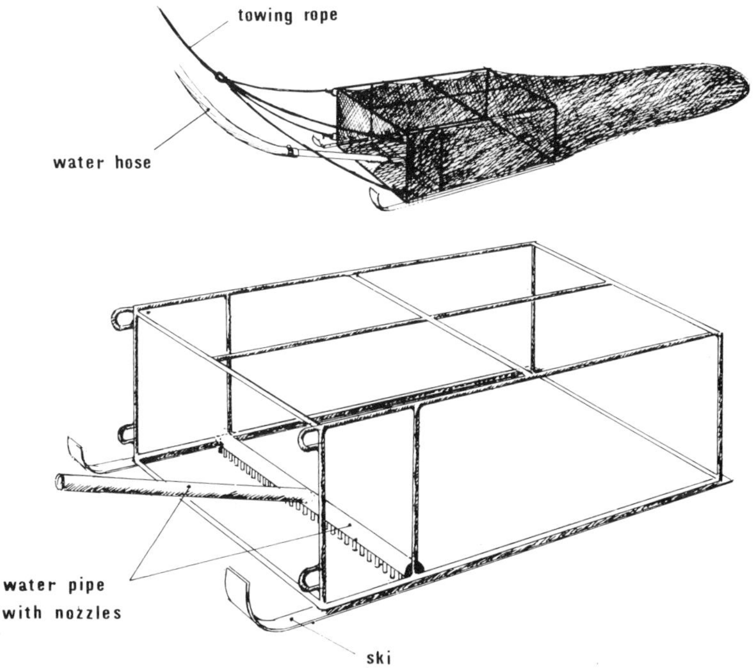

Fig. 49. Gear for collecting shrimps in the ponds, consisting of a metal frame with skis, a hose leading to a horizontal waterpipe equipped with nozzles pointing downward, and netting to collect the shrimps which jump out of the sand.

Fig. 50. The shrimps are accurately weighed in the packing shed. (Photo P. Businger.)

Fig. 51. In the packing shed, under hygienic conditions, the shrimps are placed one by one in boxes with chilled sawdust. (Photo C.R. Mock.)

B.13. The market

It is the big cities that are the market for live shrimps. Though Japan imports large quantities of frozen shrimps from various countries, this does not affect the market for the live shrimps, the Kuruma-ebi, a top delicacy, served in the best A-class restaurants only, whereas the frozen product will find its way to the B-class restaurants. The Kuruma-ebi is one of the most expensive sea food items in Japan, but the general rise in the economy of the country and the wide-spread interest in epicurism mean that the farmed shrimps are easily absorbed by the market, and the farmers can select the most profitable time of the year to deliver their product.

Upon arrival at their final destination the shrimps are revived in warm water, freed of all sawdust by careful washing, decapitated and either used in the famous "tempura" or eaten without any preparation, morsels of the flesh, virtually still alive, being dipped in soy-sauce.

C. FARMING RISKS

C.1. Hydrographic factors

As has been explained in the section dealing with hydrographic conditions in the open-air shrimp rearing ponds, the main concern of the farmer is the oxygen content of the top sediment layer and of the water layer immediately above it. Good flushing of the ponds is of paramount importance, but the administration of large quantities of feed, part of which becomes dissolved, and the production of faeces, still rich in organic material, unavoidably leads to intensive biodegradation processes requiring a considerable amount of oxygen. Therefore mechanical aerators should be placed in all the ponds, to be used continuously at night and when necessary, also during the day. Lack of oxygen and presence of poisonous hydrogen sulphide usually go hand in hand, but it is advisable to administer iron silicate if sulphide-S has to be eliminated quickly. Within a few hours iron sulphide formation leads to a drastic reduction of the sulphide-S content. Low oxygen levels and the presence of hydrogen sulphide lead to the dreaded "black-gill disease" in shrimps, usually followed by their death.

The unavoidable enrichment of the water in the ponds may also lead to excessive plankton blooms. They may be deleterious as such, especially during large scale dying-off of the plankton, but above all, what has to be feared is a bloom of *Gymnodinium* or related dinoflagellates, which produce toxic metabolites which make the shrimps suffer due to absorption of toxins by the gills. Thorough flushing is the best remedy known.

In the same category of organisms developing lavishly through enrichment of the water, the filamentous algae, which often become a serious nuisance, should be mentioned. Hand-picking is laborious and can only be a palliative. It is advised to stir and wash the bottom sediments thoroughly, at about spring tide, when the algae release their spores, making use of the water pump and the pipe with nozzles mounted on the iron frame of the shrimp net.

C.2. Predators

Birds are not considered as pests on the shrimp farms, mainly since they collect their food during daylight when the shrimps are duly buried in the sand. Fish, on the other hand, can devour a great number of shrimps. Therefore every effort should be made to keep them out of the ponds by letting all in-coming water pass through fine-meshed netting, and by using fish poisons in deeper sites, prior to stocking the pond with shrimps. Notorious shrimp predators are the red sea-bream (*Chrysophris major*), the black sea-bream (*Mylio macrocephalus*), the puffer (*Fugo rubripes*), further fish of the genera *Tridentiger, Sphyraena, Hyporhamphus, Anguilla and Sillago.* These predators will also take part of the feed destined for the shrimps. Fishes like anchovies (*Engraulis* species), silversides (*Atherina* species) and mullets (*Mugil* species) are not to be reckoned as predatory on shrimps.

If a pond is found to contain predatory fish after being stocked with shrimps it is no easy matter to remedy the situation. In Taiwan, fish farmers are said to have reached satisfactory results with tea-seed cake, poisonous for fishes, less for shrimps, but this method has not yet been tried out in Japan. A serious shrimp predator may be the penaeid shrimp itself, becoming cannibalistic when food is not offered liberally enough. Losses by cannibalism are to be feared especially during the early bottom stages.

C.3. Parasites and diseases

Contrary to experiences in farming many other vertebrate and invertebrate marine organisms, the Japanese shrimp, *Penaeus japonicus*, does not seem to suffer from parasites or diseases in the densely stocked ponds. No such losses are reported.

D. ECONOMIC ASPECTS

I. The Prefectural Seedling Centre

The Yamaguchi Prefectural Marine Products Seedling Centre, with its huge indoor installations for rearing juvenile shrimps, consisting of several roofed-in basins with a central heating system, aeration equipment and filter houses equipped with electric pumps, has no doubt cost a considerable sum of money to build and equip. It belongs to the Government, and the items with a long life are therefore not amortized in the same way as the installations of a private enterprise.

The expendible items should, however, in principle be reimbursed by the sale of seedlings to the shrimp farmers. As such can be mentioned the annual purchase of gravid females, in total about 100 kg at Y 6 000/kg. Next, an appreciable quantity of mineral oil is needed to operate the central heating system, and electricity to aerate the basins. Among the food items to be purchased, the *Artemia* eggs imported from the U.S.A. are important: 50 cans containing 0.75 of a U.S. gallon each, at Y 9 000 per can, makes Y 450 000 in total. To feed the first bottom stages, additional items such as little-neck clams and crustaceans must be purchased.

The Yamaguchi Prefectural Seedling Centre produces millions of juvenile shrimps per year, most of which are sold to the shrimp farmers of region. The remainder may be farmed to the marketable stage in the centre itself, yielding several metric tons of large shrimps. The Centre sells the seedlings to the farmer for a contract price, which includes their running costs for rearing them. In recent years the price was Y 3 per seedling. Late in the season (July—August), the price drops to about half that level.

II. The Asahi Shoji K.K. Shrimp Farm

D.1. Taxes

The Asahi Shoji farm owns the grounds on which the ponds are constructed and has to pay Y 100 000 per year in property tax.

D.2. Inventory items

The ponds of the company no doubt represent the most important inventory item. The piers and walls are partly built of masoned blocks of granite, partly of concrete blocks, so broad and solid that the company truck can reach every part. The sluice gates are of solid construction. It is difficult to express the value of these long-life items in money.

The packing shed measures 15 × 10 m and contains a freezer 10 × 12 feet. The company has plastic boats rigged as catamarans and equipped with outboard engines, motorized water pumps and special nets operating with water jets. A truck is available for transportation, e.g. to the Fukuoka airport. The company uses several aerators, cost estimated at Y 50 000 each, with a life of about 2 years. Nets of various descriptions for keeping out predatory fishes, for preventing escape of shrimps to other sections of the ponds, for catching and transporting juveniles and marketable shrimps should be reckoned among the inventory items with short life.

D.3. Expendable items

An annually-recurring item is the purchase of seedlings to stock the nursery ponds. The company does not buy the seedlings from the Yamaguchi Prefectural Seedling Centre, but from Kuruma Ebi Seisan Kumiai (a cooperative hatchery in Aeo of which the exact address is: Hiji, Aeo-Cho, Yoshiki-Gun, Yamaguchi Prefecture; the geographical position 33°59′55″N and 131°27′10″E), for a price of Y 1.5 per individual of 10—20 mg, and requires 3.2—3.3 millions of those to produce the 40 metric tons of marketable shrimps it considers as the minimum production. Food is a very important item on the budget of the shrimp farmer, counting for about 50% of the annual expenses. As such, live asari (little-neck clams) are purchased at

Y 15/kg in the shell, deep-frozen *Neomysis* at Y 40/kg and Crangonid shrimps for Y 80/kg. During the farming season, which lasts from June to December, from 6 to 30 metric tons of clams per day, plus some 2 metric tons of shrimp-like Crustaceans are used, the actual quantities being largely dependent on the supply and on the advancing size of the shrimps in the course of the season. Further, one has to count on expenses for electricity, especially to operate the aerators in the ponds, for gasoline to operate the trucks and the boats, and for packing material.

D.4. Personnel and wages

The Asahi Shoji K.K. Shrimp Farm operates on full-time basis with 20 workers and an office staff of six. In the height of the season, December, part-time workers have to be hired as well. To pack 200 kg of shrimps for the market, one needs four female and two male workers for half a day. The wages are slightly below the level paid by the industry: Y 2 500 per day. Wages represent about 25% of the production costs.

D.5. Sales

The Asahi Shoji K.K. Company sold, in the year under consideration (1973), 43 metric tons of farmed shrimps. It considers 40 metric tons a minimum, under which it is difficult to operate profitably. The prices registered for live shrimps sold at the Tokyo Central Fish Market fluctuate widely, but a fair average for shrimps of 15 g is Y 2 700/kg; 20 g, Y 3 000 kg; and 25 g, Y 3 400/kg; whereas dead but fresh shrimps make Y 1 000/kg only. The winter season, with its various feasts shows peak sales, but at the end of the season, when the shrimps are very scarce, the price may rise spectacularly. On some days in April the fantastic price of Y 9 000 per kg has been reached. From these market prices the producer has to count on a deduction of 10%.

The supply of shrimps caught naturally is so limited and declining so rapidly that it hardly affects the prices of the farmed shrimps. Trawled shrimps, moreover, make a 20% lower price than farmed shrimps. The latter being offered natural food, show an excellent flavour and are virtually undamaged, whereas the trawled shrimps may be of a less impeccable appearance.

E. GOVERNMENT ASSISTANCE

It is clear from the foregoing that the Japanese shrimp farms lean heavily on Government assistance. There are in total 14 farmers now, all in the warm southern part of Japan, but it is the Prefectural Seedling Centres which produce the seedlings to be stocked in the ponds. These centres charge a given price for the seedlings delivered, but if one considers that the Yamaguchi Prefectural Seedling Centre, covering some 22 ha, and set up in 1962—1963 at a cost of 160 million Yens, and that among all the other duties (such as nori net seeding) first priority is given to the production of seedling shrimps, one cannot believe that all Governmental expenses are refunded by the sale of seedlings.

It should not be simply concluded that it is the general tax-payer who keeps the shrimp farms going for the benefit of a small number of wealthy epicureans, for the Prefectural Seedling Centres have not only been set up to back private enterprise, but also to make an effort in the field of conservation of natural resources. The Seto Inland Sea shows a serious decline in the production of various high quality products, among which are Penaeid shrimps. Since the biological production as such has not changed, over-exploitiation must be the main cause of declining catches. On the assumption that restocking with artificially produced seedlings of valuable species of crabs, shrimps, etc. would lead to noteworthy improvement, quite a lot of money was pumped into the Prefectural Seedling Centres to remedy the precarious situation. Large numbers of seedlings of various species have already been released, but irrefutable proof that this leads to a noteworthy increase in population cannot yet be provided. Nevertheless, the fishermen are inclined to believe in such enterprises and welcome these ventures. That is probably also the reason why they gladly collect and deliver gravid female shrimps to the Seedling Centres. Experience gained around the turn of the century in northwestern Europe, in the United States and in Canada, where over-exploited waters were restocked, for example, with juvenile lobsters and cod reared ashore, has shown that too much enthusiasm combined with too little mathematical insight and too poor a knowledge of the many dangers threatening the juvenile stages in nature can amount to just throwing a lot of money into the water.

It should not be forgotten that there would in all probability be no shrimp farming in Japan were it not for the lasting enthusiasm and indefatigable endeavours of one man: Dr. M. Fujinaga, who unravelled step by step the biology of Penaeid shrimps, and found systems for rearing shrimps in the various developmental phases. Even during the war years, when he had to work on fouling problems for the navy, he could not forget his shrimps. A terrible blow was the loss of most of his notes and material when his laboratory at Aeo was destroyed by a flood caused by a typhoon. Fortunately a draft for his 1942 paper was then already in the printer's office. After the difficult postwar years he started again in 1955 in a new laboratory established at his own expense, and it was his results which finally led to the breakthrough in shrimp farming.

F. GENERAL IMPRESSION

There are now 14 shrimp farms operating in Japan, all in the warmer southern districts where it is possible to rear a crop in one season. They cover 100 ha and produce together 200 metric tons of marketable shrimps. This means a production of about 2 metric tons/ha, certainly a figure falling far short of the original expectations of a real bio-industry. The reason for this limited production is that it is virtually impossible to keep more than 250 to 300 g of shrimps/m^2 of pond bottom, even with generous feeding and adequate aeration. There have been experiments with ponds with false bottoms and air lift percolation of the sand, which does make it possible to reach a higher production, but technically and economically this does not seem to be the way to a real bio-industry.

It is the very high prices paid on the market for farmed penaeid shrimps, and the dwindling catches in nature, which make it possible and profitable to go on rearing shrimps, but there are reasons for serious concern about future developments. One expects that the Prefectural Seedling Centres will go on producing seedlings for a reasonable price, even when the tax-payer's assistance is involved, but costs for food and other expenses are rising steeply and wages are increasing steadily. It is impossible now to work profitably for a market price lower than Y 2 000 per kg, even when a good tonnage is produced. There is a feverish search being carried out for new systems able to be operated on a large scale; attempts have been made to abandon the elaborate and costly system using the specially reared *Skeletonema* and *Artemia* nauplii, for which the eggs have to be imported at a high price. Fertilizing larger bodies of water can lead to a cheaper production of food of a more natural assortment for the developing larvae, but at the same time it becomes increasingly difficult to obtain consistent results; to make sure that exactly the right type of plankton will develop for the shrimp larvae.

It will not be easy to establish more shrimp farms, or even to enlarge the present farms, for it is hard to find suitable sites, far away from industrial pollution; Japanese industry spreads above all, along the coastal areas. Moreover, the price of land is rising so steeply that it will hardly be possible to use it for a rather extensive type of farming producing only 2 metric tons/ha shrimps. The large quantities of natural food to be purchased for the shrimp farms forms another reason for concern. It becomes increasingly difficult to buy adequate supplies of live little-neck clams, and these clams are excellent and palatable food for men. In countries like Spain, Portugal, France and Italy, one would certainly pay a much higher price for fair sized specimens of this clam than the Japanese shrimp farmer has to pay today. This is not only a financial problem, but as Shigeno rightly remarks, increase in production of shrimps, which require for their growth considerably more than 10 times the amount of animal protein their own body consists of, leads to an overall decrease of animal protein on the Japanese market. Therefore, such a development cannot serve the purpose of providing the Japanese population, already short in animal protein, the type of food it needs in adequate quantities.

A way out of this dilemma would be switching over to formula food, not made from food fit for humans. The coprophagous tendencies of penaeid shrimps indicate that it must be possible to make them accept small pellets. Experience gained in the United States indicates that *Penaeus aztecus* will thrive on small size pellets (1 × 1 mm) made from chicken feather meal (25%), fish meal (15%), soy meal (15%) and shrimp peelings (30%), enriched with vitamins ("chicken starter mixture") and with 15% menhaden stick-water as binder. Switching over to pellet feeding seems to be a necessity for the Japanese shrimp farmer in the near future. Another item for further biological and technical research is the speeding up of growth. Laboratory experiments have shown that juvenile shrimps of 0.5 g can reach 20 g in 3 months if reared at 28° C, whereas at 25° C they require 4 months. This made Fujinaga conclude that shrimp farming would perhaps be more profitable in warmer countries where "abundant solar energy and warm water are really gifts of heaven". If one thinks in such terms, an increase in the amount of natural food could be hoped for by simply fertilizing the water, without administering expensive animal protein suitable for human consumption or pelleted formula feed. It seems doubtful, however, whether one can guide the processes involved in such a way that the shrimps will get plenty of suitable food but that predators and competitors stay away. Temperature as such would probably allow the rearing of 2 crops in one year in the tropics. But whether such a development will "greatly contribute towards the increased supply of animal protein to the human race", with which words Fujinaga concludes his 1969 paper, seems questionable. Items of the third and higher links in the food chain may be economically interesting to farm if they make a high price on the luxury market, but they will never be grown in such quantities and for such a price that they can serve to check the hunger of the millions. Fujinaga rightly remarks that one should not expect that the methods of shrimp farming presently practised in Japan could be commercially profitable in other countries in the temperate zone. Outside Japan, few would be willing to pay the exhorbitant price now required for farming on remunerative basis. Efforts to speed up the growth of the shrimps might lead to a greater production than the modest 2 metric tons/ha recorded now. Shrimps are nocturnal feeders and the passage of the food through the intestinal tract is quite rapid, leading to partial digestion only. In dense populations coprophagy may be a habit leading to greater efficiency and to a better conversion rate. Would it not be possible to increase the growth rate of the shrimps by making the night, during which they are active, artificially longer? It would be worthwhile to try this out on a laboratory scale first, and in case of a positive result, to consider the possibilities of technical application. Under tropical conditions this may already now be a factor leading to better growth, since the summer nights are rather short in the temperate zone.

G. BIBLIOGRAPHY

Fujinaga, M., 1962. Culture of Kuruma shrimp, *Penaeus japonicus.* Indo-Pacific Fish. Counc., Proc. 10th Sess. (Seoul, 1962), Techn. Paper 59, 2 pp.

Fujinaga, M., 1969. Kuruma shrimp (*Penaeus japonica*) cultivation in Japan. F.A.O. Fish. Rep. 57, Vol. 3, pp. 811—832.

Hudinaga, M., 1935. The study of *Penaeus.* I. The development of *Penaeus japonicus* Bate. Rep. Hayamoto Fish. Res. Lab., 1(1): 1—51.

Hudinaga, M., 1942. Reproduction, development and rearing of *Penaeus japonicus* Bate. Jpn. J. Zool., 10(2): 305—393.

Hudinaga, M. and Kittaka, J., 1966. Studies on food and growth of larval stage of a prawn, *Penaeus japonicus,* with reference to the application to practical mass culture. Inf. Bull. Planktol. Jpn. 13: 83—94.

Hudinaga, M. and Kittaka, J., 1967. The large scale production of the young Kuruma prawn, *Penaeus japonicus* Bate. Inf. Bull. Planktol. Jap., Commemorative issue for Dr. Y. Matsue, 35—46.

Hudinaga, M. and Miyamura, M., 1962. Breeding of Kuruma prawn (*Penaeus japonicus* Bate). J. Oceanogr. Soc. Jpn., 20: 694—706.

Shigeno, K., 1970. Problems of prawn culture in Japan. Overseas Tech. Coop. Agency, Tokyo, July 1970, 37 pp.

Chapter 5

EEL FARMING IN JAPAN

Example
Yamazaki eel farm, Yamashige K.K., Yuto-cho, Yamazaki, Hamana-Gun.
Director-proprietor: Mr. Mieo Miyazaki
Period of observation: 1972.

A. BACKGROUND

A.1. General principles

Eels have been a very popular luxury food in Japan since time immemorial. Especially during the summer season, the demand for eels is great, with a noteworthy peak during the Ushinohi feast, when it is traditional to eat eels to gather strength for the hot weather in the coming months. However, a fairly high price must be paid for the eel, since the catches made in rivers and lakes are always limited. In recent decades the catches of "wild" eels fluctuated between 1 700 and 3 400 metric tons per year, with a tendency to decline, since water pollution and barrage construction have led to a general decrease in water areas suitable for eels.

Like in many a place in Europe, efforts have been made in Japan to raise the production of eels by stocking, in closed water bodies, with elvers caught in the estuaries, but this did not lead to a noteworthy increase in the annual catches, and could hardly be reckoned among the cases of true farming in the aquatic environment.

A break-through came in about 1960. After experimenting in ponds densely stocked with elvers or eels, it was learned how to control the water quality, so easily contaminated by the eels, notorious waste makers that they are. In this development the Governmental Fisheries Experimental Stations played an important part. But it was a formula food, especially developed for eels, which made it possible to create a real bio-industry with a remarkably high and steady production.

Since then, eel farming has expanded rapidly. It has become the most productive sector of fresh water fish farming in Japan, accounting for 63% of its proceeds. Most of the farms are located in the central part of the Pacific coast of the isle of Honshu, especially in Shizuoka Prefecture, about halfway between Tokyo and Osaka. To some extent, one also finds eel farms in the northeastern tip of the isle of Shikoku.

Eel farming, though nowadays a real bio-industry, independent of the productivity of the water, should in reality be considered as a semi-culture, since one is still fully dependent on nature for procuring the fry. It is the elvers, born at great depth in the ocean and which migrate into the fresh water, to be caught in the estuaries during that journey, which are transferred to the ponds of the farmers. The latter observe disquieting fluctuations in the number of elvers to be caught from year to year.

Still, one needs a solid economic basis for one's farm, which means a certain minimum number of elvers to start with, year after year. It soon turned out to be impossible to collect the required number of elvers in the Japanese estuaries every year. Therefore life elvers were imported, first from neighbouring countries such as South Korea and Taiwan, which supplied good quantities of elvers, but nowadays the Japanese eel farmers even buy elvers caught in the European estuaries — from Italy, France and England. For stocking their ponds the Japanese eel farmers need every year, from 80 to 90 metric tons of elvers, and Japan itself cannot produce more than about 50 metric tons. In order of importance, the additional consignments come nowadays from Taiwan, France, South Korea, United Kingdom and Italy, and smaller quantities are bought in mainland China, the Philippines, New Zealand, etc.

The elvers are densely stocked in shallow earth ponds filled with fresh water. First they receive natural food, and later formula food, often mixed with some minced fresh fish. In Southern Japan, with its high water temperatures, growth is fast. Individually the growth varies a great deal. Some of the eels reach the marketable size in 2 years, others not before they have been fed for 3 years. Acquisition of fry, maintaining a good water quality, and control of parasites and diseases are the main worries of the Japanese eel farmer. The high price the eels make on the Japanese market renders it possible to go to considerable expense for purchase of elvers, for feed, and when necessary for disease control.

The total Japanese production of marketable eels fluctuates nowadays around 20 000 metric tons, 90% of which come from the eel farms. Most of the eel farms are private enterprises (in 1969, 1 371 out of the 1 682 farms were privately owned). The farms vary in pond acreage from 3 000 to 30 000 m^2. An average farm counts 4.3 ponds with an average area of 3 061 m^2 per pond. It is somewhat arbitrary whether eel farming in fresh water ponds should find place in a book dealing with farming the sea, but since the eel is truly a marine fish, reproducing in the sea and only temporarily sojourning in fresh water, it is felt that its incorporation is justified.

A.2. Biology of the species

The Japanese eel, *Anguilla japonica* Temminck and Schlegel, closely resembles the European eel, not only in appearance, but also in its physiological characteristics, in its behaviour and even in its flavour after being cooked. Much of what has been written on the biology of the European eel holds good for the Japanese eel. The synopsis composed by Dr. C.L. Deelder is an excellent guide in this field, and so is Dr. Tesch's recent book "Der Aal". Eels are marine fish which have their spawning grounds in the dark depths of the ocean. The eel larvae, like a willow leaf in appearance, called Leptocephali, slowly make their way to the continents, but on arrival at the continental slope they metamorphose into a glass-eel, now eel-like in shape, and but for the eyes, almost completely transparent. They migrate into the estuaries to gain access to fresh water bodies, where they hope to find an abundance of food. In the transition area, the glass-eels may be found to linger a while before they enter the truly fresh water. Then they rapidly acquire a dark pigmentation on the back side, making them in all respects really eel-like in appearance. Some migrate far into the interior, others less far, but eventually they will return to the ocean for spawning, not to return thereafter. A noteworthy metamorphosis into silver-eels characterizes the eels approaching sexual maturation and preparing for the long voyage.

Though eels of various species may be found in many parts of the world, each with its own favourite spawning grounds somewhere deep in the ocean, there are indications that they are primarily denizens of tropical or subtropical waters. Noteworthy is that they grow best in warm waters and that they are semi-dormant when water temperatures are too low. This means that growth stagnates during the winter season. For the farmer every day without growth means loss of money. Therefore eel farming has the best economical prospects in areas where water temperature is always, or at least for many months in the year, above the level at which feeding or digestion come to a virtual standstill. Japanese scientists have recorded that the Japanese eel will not grow perceptibly at temperatures under 15°C. It may take up some food, but evidently its digestive processes require higher temperatures to be effective.

A further characteristic of the eel, of considerable importance in farming, is that the individual growth varies amazingly. Starting with elvers of practically the same size, certain individuals grow rapidly, while others stay behind. Within a short time there will be a wide range of sizes. This would not be a matter of concern were it not that the eels display cannibalistic tendencies: large eels will eat the smaller ones, especially when they encounter them easily in a dense population. If one aims at a fair survival rate, one should repeatedly take the whole population out of the water to grade the eels and restock those of similar size in a clean pond. This should be performed 5 or 6 times during the whole farming period.

Another noteworthy aspect of the biology of the eel is that its sex determination apparently comes rather late in its development. Though much scientific work has still to be carried out in this field, there is sufficient evidence to assume that the glass-eels, upon arrival in the fresh water, still have the potential to become either a male or a female eel. Japanese eel biologists claim that most of the eels harvested from the ponds can be recognized as females in early maturation phases. This would certainly be most improbable if elvers were already male or female. Since female eels grow longer and larger, perhaps also faster than male, this phenomenon is of considerable practical importance.*

It is not surprising that a fish like the eel may suffer from a variety of parasites and diseases. The denser a population, the greater the risk that attacks of diseases and parasites lead to catastrophic mass mortalities. Scientific research in this field is a necessity to back the farmers in their efforts to gain consistent results.

A.3. Geographical situation

In Shizuoka Prefecture, the main area of eel farming, there is a concentration of farms around Lake Hamana. This lake is joined to the sea by a canal through the narrow strip of land originally separating it from the sea. It now contains sea water of high salinity, and is under the influence of the tides, which are of a very modest range here. On the shores of the lake, a great number of ponds (over 400) have been constructed, duly cut off from the saline water of the bay, and filled with fresh water. Most of these ponds are in use for eel farming. The firm under consideration has its headquarters on the southeastern coast of Lake Hamana, at the exact geographical position 34°42′30″N and 137°37′15″E, 10 km west of the city of Hamamatsu. The farm has 1.7 ha of ponds at its headquarters, and in addition, about 3 ha of ponds in the southeastern-most corner of Lake Hamana at 34°41′20″N and 137°39′15″E.

A.4. Hydrographic pattern

The hydrography of Lake Hamana, with its deeper and shallower part, showing respectively, muddy and sandy bottoms, is not of importance for the eel farms, since the rearing ponds contain fresh water. The elvers, coming from the ocean, pass Lake Hamana on their way to fresher water and it is especially in the narrows, leading to the northern section of the lake, that one can profitably place nets for catching elvers.

The ponds. The Yamazaki farm operates with 10 outdoor ponds, and since 1970 with a 2 000 m^2 elver rearing house (Fig. 53). The size of the outdoor ponds ranges from 3 000 to 5 000 m^2. The ponds are 70 to 100 cm deep, all with a corner site up to 1.50 m deep for fishing out. The bottoms of the ponds are of a sandy mud, also in the rearing house. The walls of all the ponds are built of masoned concrete block, impossible for the eels to dig into.

The 2 000 m^2 of the rearing house holds 3 larger basins of 400 m^2 each,

* When the Japanese ponds are stocked with elvers purchased in Europe, the percentage of males is said to predominate in due course.

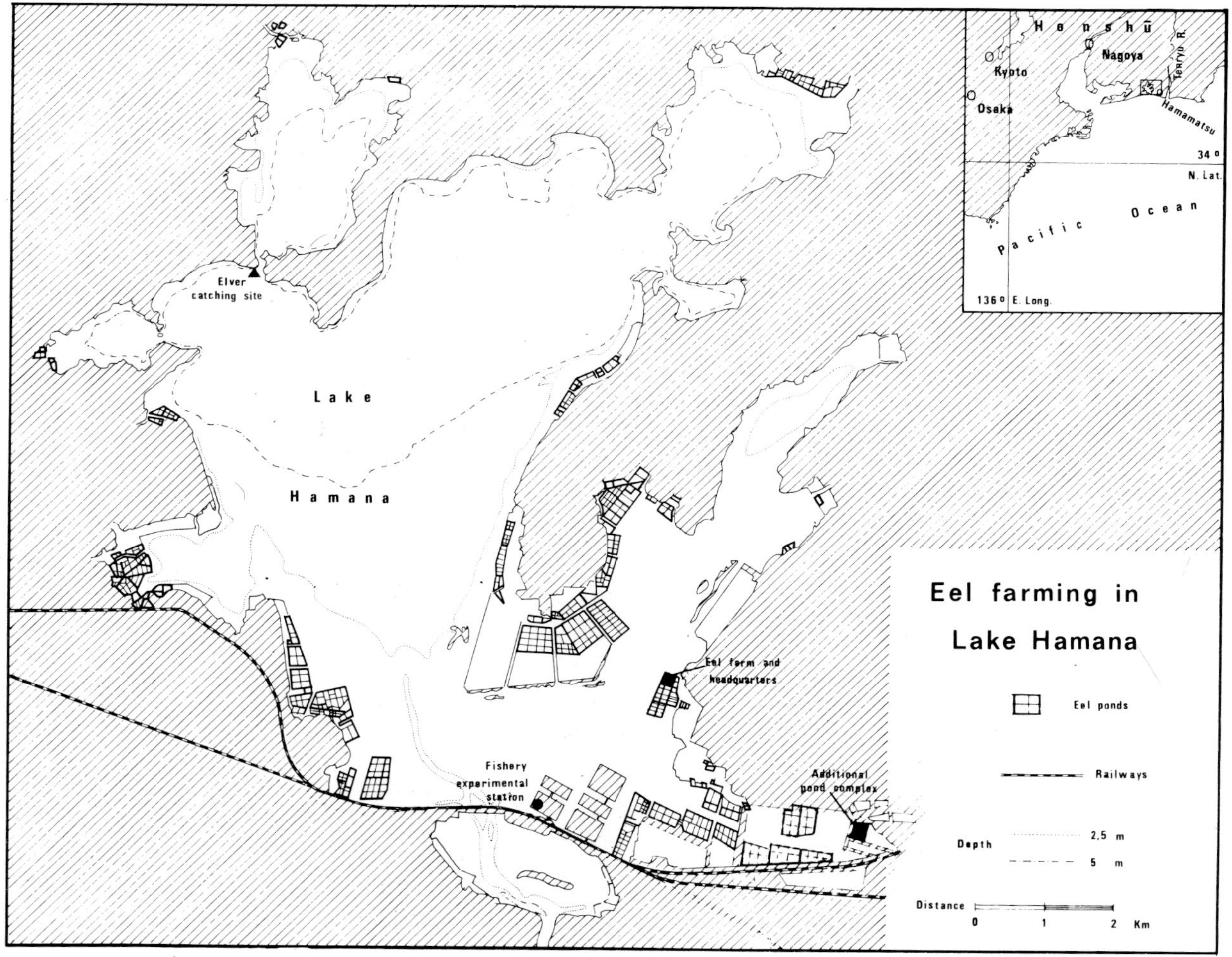
Eel farming in
Lake Hamana
Eel ponds
Railways
Depth
2,5 m
5 m
Distance
0
1
2 Km
Honshū
Kyoto
Osaka
Nagoya
Tenryu R.
Hamamatsu
34°
N. Lat.
136° E. Long.
Pacific Ocean
Lake
Hamana
Elver catching site
Eel farm and headquarters
Fishery experimental station
Additional pond complex

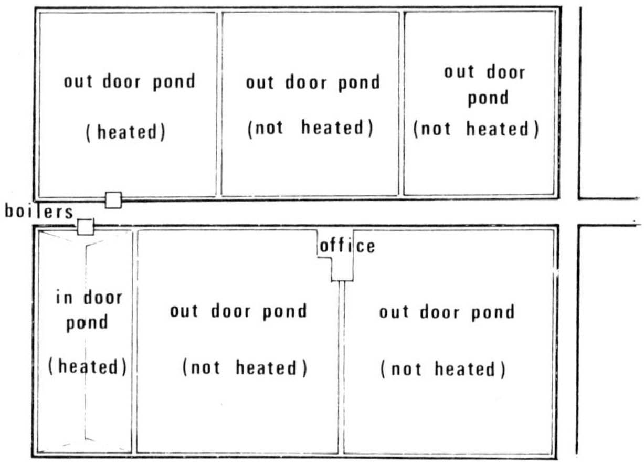

Fig. 52. Lay-out of the ponds of the Yamazaki eel farm at the headquarters.

Fig. 53. The interior of the rearing house of the Yamazaki eel farm. Note the long-shafted paddle wheel, the feeding box, the wide plastic tubes for control of air temperature and the thermostat dangling from the ceiling. (Photo Y. Akimoto.)

and 4 smaller basins of approximately 60 m^2 each. These basins have depths of 100 and 60 cm respectively.

The source of water for the ponds. All the ponds should contain fresh water, and a certain degree of flushing is a prerequisite in eel farming. In principle, all the water in each pond should, at least in the summer season, be completely replaced by new water in the course of one month. In summer fresh water is brought into all the ponds every night. Large volumes of water have to be available. The main source of fresh water is the Tenryu river, 20 km further west. A pipeline of large diameter (100 cm at the source) guides the Tenryu river water towards Lake Hamana to serve a great number of farms. Three inch pipes bring the river water to the ponds. A second source of fresh water is needed in summer, when the supply of river water declines. Phreatic water is pumped up from wells drilled to a depth of 150 m, in which artesian water rises to about 7 m below the surface. Use is made of electric under-water pumps of 15 to 30 h.p.

Water temperature. The ideal water temperature for eel farming is about 25° C, but this level is only reached and even surpassed in the warmer times of the year, to drop below the biologically critical level of 15° C towards the end of the year. In winter, the water temperature drops still further in the rather shallow ponds, occasionally even down to the freezing point. One has to wait to some time in April to see the water temperature rise again to 15° C, critical for eel growth. In the open ponds, eels will therefore be in a semi-dormant stage during the colder months. To avoid retardation of growth of the young eels, a rearing house has been constructed in the Yamazaki eel farm and has been in use since the year 1970. It contains basins of different sizes and is roofed with a steel and glass superstructure, covering 2 000 m^2 in total. The water in the ponds can be kept at 20—25° C by an oil central-heating system, leading hot water through pipes installed on the bottom of the basins. In addition, under the whole length of the roof there are wide plastic tubes with a series of small openings to blow hot air (heated electrically) through; a centrally placed thermostat controls the air temperature. This combination of water and air heating operates economically, the warm air with its low specific heat serving as insulator. One should take into consideration that the indoor basins too have to be flushed regularly, cold water being led in. One of the smaller outdoor basins, measuring about 3 000 m^2, is also equipped with a tubing system on the bottom, connected to the central heating (Fig. 54). It would, however, lead to prohibitive costs if the water temperature in this outdoor pond were kept at 22—25° C throughout the winter season. This outdoor basin is only heated when the air temperature is already quite high, so that it will not be too costly to keep the water temperature at 22—25° C. In practice, this heated outdoor pond is only used when the basins of the rearing house threaten to become overcrowded in

Fig. 54. The heated outdoor pond of the Yamazaki eel farm. (Photo P. Businger.)

May or June. Shortly after, the central heating can be switched off completely.

Chemical composition. The chemical composition of the water in the eel ponds should be kept under careful control. In doing so, special attention should be paid to the colour of the water. In cases of doubt, the farmers bring water samples to the nearby Fisheries Experimental Station, where they get advice free of charge. The pH of the water should by preference be kept between 7.8 and 9.0. When one has to use phreatic water of low pH, addition of lime is a necessity. As a rule there is a pH of 8.0—9.2 during the day, dropping to 6.8—7.2 at night.

Due to the dense eel population and the heavy feeding, poisonous constituents such as nitrite and ammonia may rise to a dangerous level. The nitrite content should be kept under 0.2 p.p.m. and the ammonia under 2 p.p.m. Flushing helps to prevent dangerously high levels.

Oxygen. One of the main headaches of the eel farmer is keeping the oxygen content of the water at the desired level. During the night, the phytoplankton interrupts its production of oxygen and shares oxygen consumption with the eels. But also during the light hours of the day, it may be that the phytoplankton population is so dense that insufficient light (less than 200 lux) penetrates to the deeper water layers to make oxygen production possible. When the eels observe that the oxygen tension becomes too low in the deeper layers of the ponds, either during the day or at night, they start swimming near the surface to avoid asphyxiation. Practical experience is that the eels will not grow under such conditions. To promote aeration and vertical mixing of the water, so that the eels will find favourable conditions at any depth in the ponds, floating aeration and agitation apparatus has been installed. One widely used device consists of an axle, kept in motion by an electric motor, bearing at both ends a paddle wheel. Floats bear-up the construction and the engine is duly protected against splash in a sheet steel housing. There are numerous brands of such aerators, and some firms make use of electrically-driven screw jacks to keep the water well aerated.

In the larger ponds aerators with short axles are used, in the smaller indoor ponds those with longer axles, protruding about 2 m at each side. In the smallest indoor basins, the aerators are supplemented: to aerate the water still better, air is blown in through a series of perforated plastic tubes.

Fig. 55. The short-shafted paddle wheels used in the open air ponds are indispensable in eel farming. (Photo Y. Akimoto.)

Fig. 56. In other ponds screw-jacks are used to mix and aerate the water thoroughly. (Photo P. Korringa.)

Phytoplankton. The water of the ponds is dark green, caused by a dense vegetation of small-celled phytoplankton organisms. The transparency of the water is limited, and one cannot see the bottom; occasional dead eels can escape the attention and decay on the spot, consuming much oxygen. The dense plankton vegetation is not necessary for the eels but is an inevitable consequence of their dense population and of the great quantity of organic food administered to the eels; they are notorious waste makers. Biodegradation processes produce the nutrients for the plankton, which in its turn may deteriorate the water quality by becoming too dense. Flushing of the ponds is the best remedy. A good indication of the content of dead organic matter is the C.O.D. (Chemical Oxygen Demand), to be assessed in the laboratory with sodium permanganate. It should not surpass the 10 p.p.m.

A.5. Legal aspects

The acronimic "K.K." behind the name of the company stands for "kabushiki kaisha", which means that it is a limited enterprise. The eel farm under consideration is, like almost all the other Japanese eel farms, a private enterprise, run by the family. The owner is assisted by his wife, part-time also by his two children. In addition there are two hired hands. It takes about one full-time hand to handle 2 ha of eel farm over a year.

Normally the land on which the ponds are constructed is owned by the farmers. The farmer is a member of a cooperative which handles the marketing of the eels, the usual pattern in Japan.

B. SEQUENCE OF OPERATIONS

B.1. Preparing the ponds

It is advisable to empty a pond completely and to clean it before restocking. A pond should be filled with new fresh water 2 to 3 days before stocking. If a basin of the rearing house to lodge juvenile eels previously grown in heated water is to be filled, the temperature of the new water should be raised prior to stocking. The same scheme holds good prior to making use of the heated outdoor pond.

B.2. Catching the elvers

The first elvers arrive in the estuarine water of Shizuoka Prefecture in December. The elver run may last until May, and usually has its height in February and March. It is not the eel-farmers themselves who catch the elvers required for stocking their ponds, but rather individual fishermen operating with different gear. In Lake Hamana there are about 2 000 persons engaged in elver catching, most of them operating with simple scoop nets, sometimes combined with a faint electric light to lure the elvers. About 100 fishermen operate with a stationary net, placed in well-selected sites with a depth up to 2 or 2.5 m. Between two cedar poles of about 7 m length, placed 20 m apart, is stretched a net of 40—50 m length, reaching from the surface to the bottom. The net, made of knotless nylon, 10 meshes per inch, gradually tapers towards a small meshed (16 mesh/inch) cod-end, which is strengthened with a plastic hoop and provided with a funnel to prevent the elvers from swimming back. The cod-end is kept in position with a third pole, and a floating light is placed over the trap portion of the net to lure the elvers. The nets are set out at slack water and during the run of the tide the fisherman comes 3 to 4 times with his small wooden boat (1—2 tons, usually equipped with a 10—20 h.p. gasoline engine) to collect the catch. During the flood tide, the elvers are said to swim in the higher water layer, during the ebb tide they try to make their way inland along the bottom. This behaviour facilitates their migration towards the fresh water bodies. The best catches are made during moonless nights, when a gentle breeze (1—2 knots) blows. Daily catches of 500—600 g are considered as normal.

The Japanese elvers are, upon arrival at the coast, somewhat smaller and

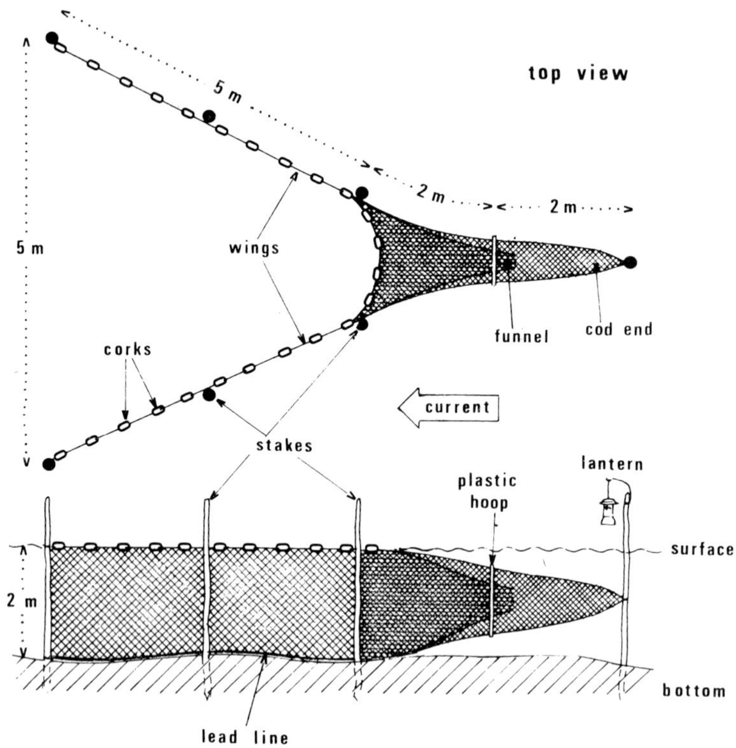

Fig. 57. Net for catching elvers in top-view and side-view.

lighter than those collected in Europe: they measure 5.3—5.7 cm and 5 000 in number per kg, whereas the elvers caught in the Netherlands measure 7.0—7.5 cm, 2 500 per kg; those from the coast of France even 8.0 cm and 2 000 per kg. The elvers caught at the Japanese coast are still truly glass-eels, lacking in pigmentation, as their name "shirako", meaning "pale", indicates.

B.3. The purchase of elvers

The total catch made in Lake Hamana fluctuates between 1 to 7 metric tons per season, with an average of about 3 metric tons. The season 1971—1972 was poor: 1.5 metric tons were caught. The price paid by the dealers to the individual fisherman varies with the supply from Y 0.50 to Y 10 for each elver.

The eel farmers of Lake Hamana need some 30 metric tons of elvers each year to stock their ponds. They buy them from brokers. It is clear that the local catches should be supplemented. Other fishing sites in the Shizuoka

Fig. 58. Elver nets placed in Lake Hamana. Note cables leading to the lamps. (Photo P. Businger.)

Prefecture supply some 5—6 metric tons, sold on a quota basis, the remaining tonnage is imported, mostly from Korea, where the elvers are still smaller than in Japan, and from France. The dealers can keep the elvers alive in floating boxes, but also out of water, provided the humidity in the containers is high. The Yamazaki farm buys 30 to 100 kg of elvers (i.e. 150 000 to 500 000 specimens) per year, in batches of 5 to 10 kg, spread over the season.

B.4. Stocking the ponds

It is a remarkable fact that the elvers caught in water of high salinity can be transferred one day later to rearing ponds containing fresh water. They evidently can stand such an abrupt transition unaffected. Very soon after stocking the glass-eels develop dark pigmentation, giving them a real eel-like appearance. As is always done in fish farming, one uses the smallest ponds for the smallest specimens. When placed in a pond in which the water temperature is below 12 to 13°C, the elvers dig immediately into the bottom, not to show up before the 15°C level has been surpassed, but because the Yamazaki farm, since the year 1970, has a heated rearing house at its disposal, the elvers are invariably stocked in one of the small basins of the rearing house where the water temperature is high enough to keep them feeding actively from the very beginning. The stocking rate is 100—300 g per m^2. For Japanese elvers this means about 1 000, for European elvers about 500 per m^2.

B.5. Aeration, turbulence, flushing and heating

In the heated ponds of the rearing house, aeration and agitation is necessary from the moment the elvers are put in, but in the non-heated open air ponds, the aerator-agitators are not needed before June, when the water temperature rises. The aerators are preferably used at night, but when necessary continuously. At night the phytoplankton consumes oxygen rather than producing it, which makes aeration all the more necessary, and moreover, the rate for electricity is considerably lower at night. Good aeration should lead to normal behaviour and active feeding of the eels.

During 3 or 4 h of each night, it is advisable that new fresh water be brought into the rearing ponds, to replace about 4% of the volume; when the water temperature is 15°C or higher, this is compulsory. Similarly in the heated ponds of the rearing house, even though it takes an appreciable amount of energy to heat this new water to the desired 22 to 25°C level. When the water quality deteriorates, or when a pond is fished out, three-quaters to four-fifths of the water should be replaced. Of course, the new water should be warmed up to 22—25°C before restocking, if it concerns an indoor pond. The heating season is from December to July in the rearing house, May and June only in the heated open air basins.

B.6. Feeding the elvers

In the non-heated open air ponds the elvers need not be fed in the first weeks after stocking. As soon as they crawl out of the bottom again, as the water temperature rises, they may find some natural food, such as Chironomid larvae and Daphnia's, but it is advisable to supplement this with other feed. In the rearing house they have to be fed from the very beginning. So in the first week "mimizu", oligochaete worms such as Tubifex and Limnodrilus gotoi are used. This food is bought from specialized dealers, also serving aquarium holders, who purchase it in turn, from those who collect it in somewhat polluted muddy waters.

In the beginning, the elvers are fed several times during the night, use being made of a faint electric light to attract the elvers. After having been fed for one week on "mimizu" the elvers receive minced fresh mackerel, to which is added, from the beginning, some 10% formula feed (see Technical aspects). Of "mimizu", the elvers get 5—8% of their body weight per day, and when they have been switched over to mackerel they should receive 20—30% of their body weight per day, in three rations. Gradually the percentage of formula food is increased so that the ration finally becomes 80% formula food and 20% minced frozen mackerel. When the water temperature is over 20°C, the eels are fed about 12—13% of their body weight of this compound, at 15 to 20°C, half that dose. Both in the open air ponds and in the rearing house, the food is administered in special boxes, keeping out the daylight. There the eels find food in wire-meshed trays, in a site dark enough to be used during the day. When they are used to formula food, the eels can be given their feed once a day — early in the morning.

The switch-over from Tubifex, via minced mackerel, to formula feed offers little difficulties in the heated indoor basins, but in the non-heated open air ponds it is often not an easy matter to get the juvenile eels, just emerged from the bottom, used to non-living feed.

B.7. Growth and grading

The elvers grow fast when they get plenty of food. For two reasons they cannot be kept long in the basin in which they were put: firstly because, as

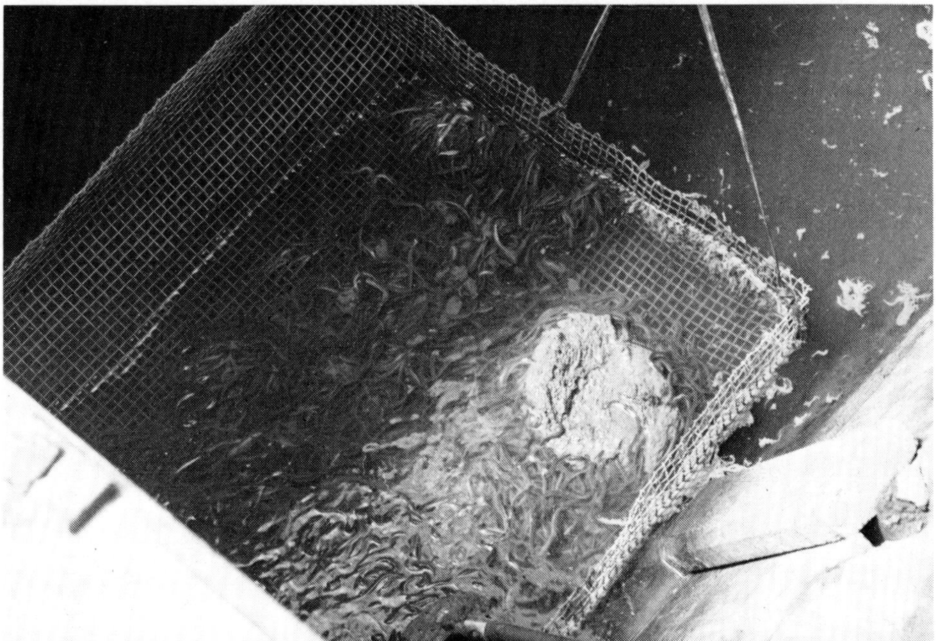

Fig. 59. Small eels feeding on dough of formula food offered in a tray of wire netting placed in the dark feeding box. (Photo P. Businger.)

the total weight of the eels increases, more and more oxygen is consumed, and from a certain point onwards it becomes difficult to supply the eels with sufficient oxygen with the aerator. Secondly, the individual growth varies widely, and too great a size range leads to losses through cannibalism. Thirty days after stocking the pond with elvers, selection for transfer begins. Juvenile eels are caught when they are congregating during feeding with a simple scoop net of about 40 cm diameter. To grade them a steel grid with adjustable slits is used. The elvers passing through the slits can remain in the pond, those held back are transferred to another. This procedure is repeated once a fortnight until all the young eels are transferred to other basins, usually some time in July. Outdoor ponds are put in use as soon as the water temperature allows it. If it is decided to fish out a rearing-house pond, for instance to change the water completely, one uses a nylon net fastened to a steel frame (Fig. 60) bearing two steel prolongations to be pushed into the bottom. The deeper corner of the basin is cut off obliquely with planking and the net is placed in the central doorway. The water is pumped out from the cut-off corner, or simply drained off by gravity when this is possible, and gradually the young eels will come to the net as the water recedes.

In this first rearing phase the elvers will reach 5 to 10 times their original weight. Some losses are inevitable: about 30 to 40% was considered as "normal" where the open air ponds were used, but in the rearing houses the survival rate may be much greater, up to 95%.

B.8. Repeated grading and transfer

For the same reasons explained above, the grading and transfer of the growing eels must be repeated also when they are kept in larger outdoor basins. Between the elver stage and the marketable stage all the eels are transferred 3 to 5 times to other ponds. Graded eels can be restocked rather densely: 300 to 1 000 g per m^2, irrespective of their individual size.

For catching the eels to be graded from the outdoor ponds, the same principle is used as applied for the juveniles in the indoor ponds: when they congregate at the feeding site they can easily be caught with a net. This net is made of strong nylon twine and has meshes of 4—5 mm.

When the eels have reached a weight of 10 g, a special tool is used to grade them. It consists of a wooden frame, kept in an inclined position, of which the bottom of the centre part is made of steel bars placed lengthwise. The width of the slits between the bars can be adjusted. The eels are dumped in a receptacle at the top and are allowed to crawl down over the moist grid. Those which can pass through the slits drop into a basket placed under the grid; the larger ones crawl down over the whole length of the grid and are collected in another basket, placed at the lower end. This system of grading is used throughout the whole farming procedure. The eels are invariably restocked in new ponds at a rate of 300—1 000 g per m^2.

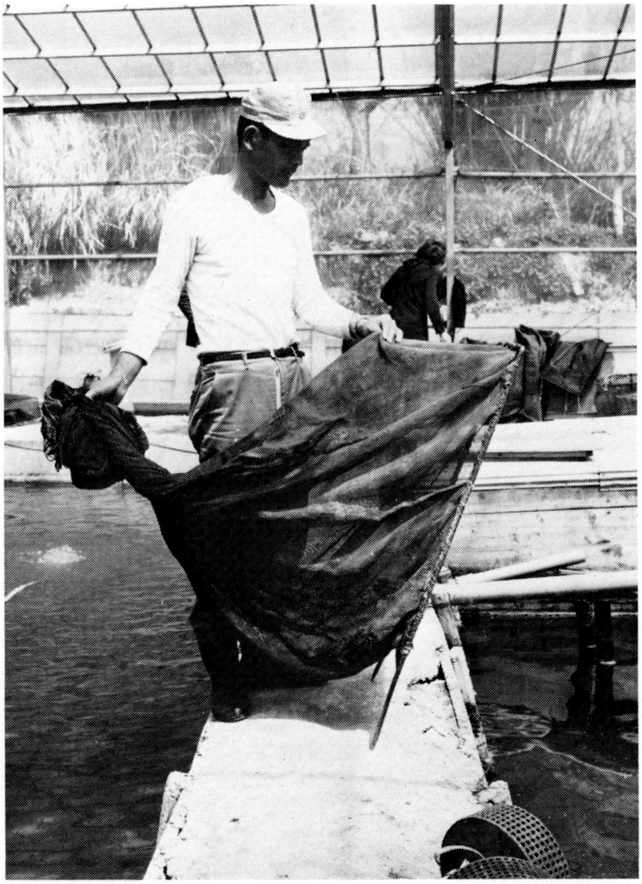

Fig. 60. Nylon net on metal frame for fishing out indoor elver ponds. (Photo P. Businger.)

B.9. Feeding and growth

During July all the eels will have been transferred to the larger ponds. They receive their daily dose of feed in the early morning but can come to the darkened feeding boxes whenever they wish. The formula food is offered as a big lump of dough: to the dry meal-like substance water is added (110 parts of water: 100 parts of meal), and it is kneaded into a dough in a type of motorized kneading machine such as bakers use. Some fish oil (5—10%) and some minced fish (up to 20%, e.g. deep-frozen mackerel) should also be administered. At 80% formula meal, the feeding rate remains at 12—13% of the body weight per day at a water temperature of 20°C or more. The feeding quotient is rather high: 7.5—9; higher for older eels than for the juveniles. On basis of the dry formula food it is 1.2—2.6, i.e. about one-third or

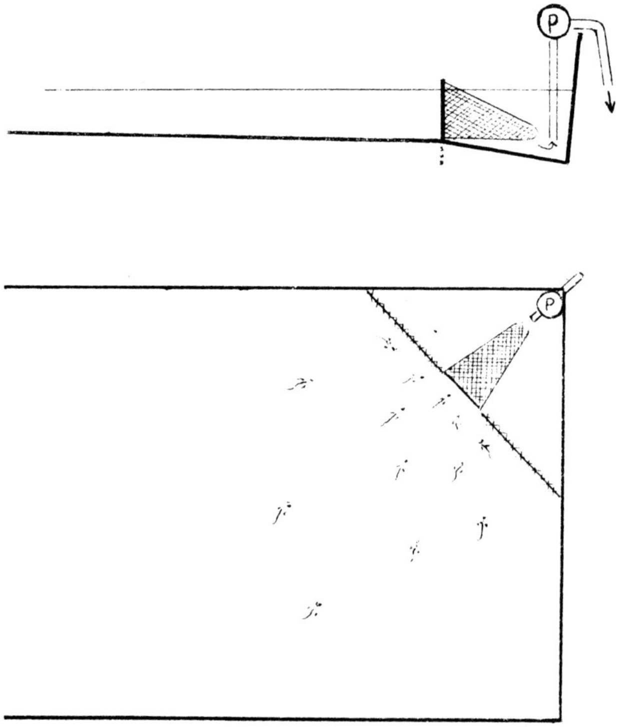

Fig. 61. Technique for fishing out indoor elver ponds; P = pump.

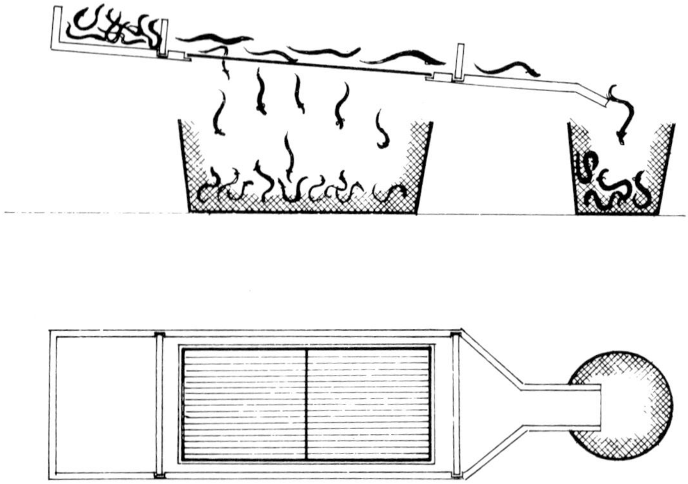

Fig. 62. Apparatus for grading eels in side-view and top-view.

Fig. 63. The feeding house straddling two ponds of the Yamazaki eel farm gives the eels ample opportunity to take their feed at any hour of the day. (Photo Y. Akimoto.)

one-quater of the fresh fish basis. One counts on the production of 0.7 kg of eel-flesh per kg of dry formula food. From an individual weight of 10 g onwards, the losses suffered are rather limited if no diseases or parasites strike: some 10% per year. One can count on 70% survival of the young eels stocked in the ponds.

B.10. Harvesting the marketable eels

The individual growth rate remains remarkably different, and the most aggressive feeders and/or the most efficient digesters are the first to reach consumption size: 40—50 cm length and 150—250 g in weight. Since the Japanese market pays a lower price for bigger eels, they should be harvested before their weight surpasses 250 g. From the eels put in the larger ponds in April, the fastest growers, some 5—10% of the stock, will reach the marketable size as early as November of the same year. The others will begin to grow more slowly as winter approaches, to pass the coldest months in a semi-dormant stage, resuming feeding and growth when the water temperature surpasses 15°C again. Most of the eels are ready for the market in the course of their second year of sojourn in the outdoor ponds. Mid-summer (July) shows a peak in the sale of eels.

Again, the most convenient way to catch eels is by net, when they congregate at the feeding site. Occasionally one has to remove the fish out of a whole pond. The large eels are graded in the way described above and classified according to weight.

B.11. Marketing and consumption of eels

The live eels are shipped to the market in batches of 10 kg, packed in plastic bags. No water is added, just some ice to keep the temperature low, and pure oxygen. Then the eels will stay alive for 20—40 h, even in hot summer weather. For selling his product, our farmer follows the usual pattern, invoking mediation of the local sales co-operative. Favourite eel dishes are "kabayaki", consisting of spitch-cooked eels on skewers, flavoured with a mixture of soy sauce, sugar and "mirin" (rice wine); and "unagi donburi", the same without skewers, served in a bowl of hot steamed rice. The eels are not skinned before cooking in Japan, unlike in Europe, but they are invariably split lengthwise to remove the vertical column and to disguise their snake-like appearance.

C. FARMING RISKS

C.1. Hydrographic factors

The hydrographic conditions in the ponds should be kept under careful control. Good flushing, aeration and agitation are essential in eel farming. But even then the eels may start to die when the water temperature surpasses 30°C in summer, exactly in the season that the supply of fresh water may give reason for concern. It is said that eels imported from Europe are more sensitive to high water temperatures than the Japanese eels. The fact that elvers imported from Europe often show higher mortality rates than the Japanese elvers may be due to the conditions prevailing during transport. Elvers and eels under stress may produce so much slime that suffocation follows; air transport often meets with delays.

C.2. Predators

The Japanese eel farmers make no mention of losses due to predation. Their ponds are well surveyed and diving birds such as cormorants evidently do not dare to come so close. It is therefore unnecessary to string nylon wires over the ponds.

C.3. Parasites and diseases

Diseases and parasites form the major risk in eel farming, which is not surprising when one considers the great density of the eel population in the

Fig. 64. Upon advice of the Fisheries Research Station the farmer may have to spread lime in the eel ponds to correct the water quality. (Photo P. Korringa.)

ponds. More than 10 different diseases and parasites are known to attack the eels in the ponds, and were it not that the Shizuoka Prefectural Fisheries Research Station, Hamanako Branch, is always ready to advise the eel farmers free of charge on the measures to be taken to control these pests, it would hardly be possible to reach consistent results with eel farming, and to work on a profitable basis. The predominant diseases are gill disease and red fin disease. The crustacean parasite *Argulus* may be troublesome too. A list of the parasites and diseases observed in the Japanese eel farm, together with notes on methods of control, is given in the last pages of the chapter, in the section on technical aspects.

Pest and disease control costs Y 200 000 per hectare annually. In some cases, the chemicals have to be added to the water of the ponds, in other cases it is more practical to bathe the eels in a dilute disinfectant prior to restocking.

It is necessary to be alert, for it may be that diseases thus far unknown strike. In 1970, for instance, there was serious mortality among the elvers caused by branchio-nephritis, allegedly imported with elvers from Europe.

D. ECONOMIC ASPECTS

D.1. Inventory items

The ponds represent the most important capital investment of an eel farm. The outdoor ponds have a depth of 70—100 cm, the bottom consists of a muddy sand, and the walls are built of masoned concrete blocks. They have been constructed years ago, and for the amortization one can count on a very long life. Construction of an open eel pond of 2 ha would now cost Y 15—20 million. The 2 000 m^2 rearing house with central heating system was built in 1970 at a cost of about Y 5 000 000. Other inventory items are the office building plus storage sheds, the aerators, the fish mincing machine, the kneading machine used to make a dough of the formula food, the 15—30 h.p. electric pumps for pumping up phreatic water in summer, the grading apparatus and various nylon nets. The farmer does not need a net to catch elvers in Lake Hamana, since he does not participate in these operations himself. Such a net would cost, according to size and quality from Y 50 000 to 100 000 and would last two or three seasons.

The aerators, of which, at least one is needed for every pond, the indoor ponds included, cost, according to the type selected, Y 70 000—150 000 each, and have an estimated life of 3 years.

The company owns one truck which is used to transport marketable eels to a shipping point.

D.2. Expendable items

The purchase of elvers is to be booked as a recurring expense of considerable importance. This farm buys from 30 to 100 kg of elvers each year. Japanese elvers cost Y 50 000 per kg, Korean elvers Y 70 000—100 000/kg, and French elvers Y 12 000 per kg. The price difference does not reflect the quality, but rather the number of elvers contained in a kg, high for Korean elvers, much lower for French elvers. The price given for purchase of Japanese elvers is an average for the last few years, but in reality the price fluctuates widely within one and the same season, a question of supply, demand and speculation. Purchase of elvers counts for about 25% of the total expense of an eelfarm. A still more important expendable item is the feed, counting for 50—55% of the total annual expenses. Formula food is purchased in 20 kg sacks and costs Y 100—120 per kg. Frozen mackerel costs Y 30—50 per kg, and the oligochaetes used in the early phase are purchased for Y 300 to 500 per kg. The expenses for feed do not weigh heavily during the early phases of rearing: less than 5% of the total.

As fuel for the central heating, heavy oil — grade A — is used. To keep the water in the indoor pond at the desired temperature costs Y 1 500 000 per 1 000 m^2. The capacity of the heating system is 7 200 kcal/100 m^2. Calculated per eel, heating would cost Y 4.62 in fuel when the firm keeps 30 kg of elvers, but only Y 1.40 when 100 kg of elvers are stocked in the rearing house. For heating the outdoor pond in spring, costs are about the same.

Heating the air above the water sheets in the rearing house is done by electricity and requires 0.37 to 0.75 kW/8h. When the aerators are operated at night, it costs Y 50 to 80 per night per unit.

Five yen per m^3 is paid for the river water directed to the farm from the general pipeline, and in principle, one needs per month, a quantity representing the whole volume of all the ponds for flushing. For pest and disease control one should count on an expense of Y 200 000/ha.

D.3. Personnel and wages

Both the farmer and his wife work on the eel farm, and their two children part-time. Two hired male workers are also employed. They get a wage of Y 70 000 per month and work in the main season (April—October) 7 days per week and 10 hours per day. On average in Japanese eel farming, one full-time worker is needed per 2 ha of ponds.

D.4. Sales

The actual wholesale price for eels of 150—200 g each fluctuates between Y 1 300 and 1 400 per kg. For bigger eels one gets only Y 800—1 000 per kg. This is a high price compared with the European eel market and indicates the luxury character of the product. The Yamazaki eel farm produces about 75 metric tons of eel per year, viz, 15 metric tons/ha. This is a good production level, for the average farm in Shizuoka Prefecture produces 10 metric tons/ha per year and the maximum observed was 30 metric tons/ha. The gross income of the company amounts to about Y 100 million per year.

E. GOVERNMENT ASSISTANCE

The Japanese eel farmers are assisted by the Government in several ways. Of the greatest importance are the activities of the Shizuoka Prefectural Fisheries Research Station, Hamanako Branch, from which the farmers get valuable advice. Not only on control of parasites and diseases, but also on biological and technical aspects of eel farming — feeding and growth included. The Research Station itself has various ponds and is engaged in thorough studies of various aspects of eel farming. Eel farmers are, for instance, faced with a fierce competition between the producers of formula feed, and since feed constitutes a very high percentage of their annual expenses, they are inclined to economize on this item. Their own practical experience, and the advice of the scientists and technicians of the Research

Institute have, however, made them reluctant to deviate much from the quality of formula feed considered as optimal in eel farming, and has convinced them that use of a special type of formula food for the youngest phase is advisable. Through mediation of the sales cooperatives, in which the Government has a stake, one can be sure of a reasonable wholesale price, since fierce competition can thus be avoided.

The pipeline from the Tenryu river, constructed by the Prefectural Government, in co-operation with the Central Government, is an invaluable asset for the eel farmers around Lake Hamana and for various other users.

F. GENERAL IMPRESSION

Eel farming Japanese style is an interesting case of farming a marine organism, almost reaching the level of a bio-industry. Considering that the eel is a predatory fish, representing the third or the fourth link in the food chain, depending on what the main items of its diet are, the production is certainly amazing: 15 metric tons per ha per year in our case. From very productive eel water in Europe one can harvest up to 50 kg of marketable eels per ha per year, but as a rule the yield is only 15—20 kg/ha. This tremendous difference (a thousand-fold) is, of course, due to the fact that eel farming is not comparable with eel fishing in the wild, the difference being that the farmed eels are kept in as limited a volume of water as feasible, and are fed with the optimum quantity of formula feed mixed with some minced fish.

Eel farming is, however, still a semi-culture, since a very important factor, the supply of elvers, is largely dependent on what nature yields. Without importation of elvers from abroad, the production of the Japanese eel farms would drop back to an almost insignificant level. That now, some 90% of the eels demanded by the Japanese fish market comes from the farms, has been made possible through large scale importation of elvers, not only from neighbouring countries, but also from Europe. The supplies are, however, far from constant and therefore the future of eel farming in Japan is rather uncertain. If other countries start eel farming on a large scale this could lead to cutting off part of the present elver supplies for Japan. A strong factor in favour of the Japanese industry is, however, the fact that eels are a luxury product in Japan, for which a very high price is paid in comparison with other countries. Therefore the Japanese farmers will not shrink from paying still higher prices for elvers. The purchase of fry constitutes now about 25% of the annual running costs of an eel farm. Introduction of heated rearing houses not only speeds up the growth, but also reduces the mortality in the younger phases considerably. If all the farmers built such rearing houses Japan could produce the same quantity of marketable eels as today with about half the tonnage of elvers, or perhaps even less.

Since it would be of considerable importance to be independent of importation of elvers, and to convert the present way of farming into a real bio-industry, independent of the whims of nature, the possibility of rearing elvers from the eggs has been examined. Mature males or females are not found in nature, but after lengthy experimentation with hormone injections a few mature male and female eels have been produced. This is, however, only the first step in the venture. Since eel larvae (Leptocephali) grow up in the abyssal depths of the ocean, to come later to higher layers and take the long journey to the continents (in the case of the European eel even lasting years), it does not seem an easy matter to offer them optimal conditions in a shore establishment. There is still a long way to go before closed-cycle farming will be possible.

A second bottle-neck in eel farming is often the fresh water supply. For the farms around Lake Hamana, there is river water brought by a big pipeline, but even then it is necessary to pump up additional phreatic water in the summer season. When a lesser quantity of fresh water is available, the basins cannot be flushed so well, hence a smaller number of eels can be kept per m^2 of pond. Low density stocking reduces the remunerativeness of the farms. A greater water supply makes intensive flushing possible, and thereby a dense stocking and a high yield per ha. It is a question of economy to ascertain the optimal relation between flushing and stocking in the farming area under consideration.

Impressed by the high tonnage of eels produced, some biologists are inclined to advise developing countries to start such ventures themselves, to produce valuable animal protein for their own population. This is, however, an erroneous view, for eel meat is an expensive luxury item based on feeding a multiple weight of good, fresh fish protein worked up to formula feed. Eels will not accept the normal low grade fish meal in their formula feed, but demand as basis for their food, very fresh flatfish, converted into a white odourless meal. No developing country could afford to use a great tonnage of good fish protein to convert into a meal for producing a much smaller tonnage of luxury eel protein unless they would be able to sell that abroad for such a high price that they can buy, in return, a much greater quantity of good animal protein of another nature. Since the luxury market demands live eels, there is as yet little chance for developing countries at great distances from the Japanese cities to participate in eel farming.

G. TECHNICAL ASPECTS

G.1. Formula food

The formula food for eel farming, now produced by about 10 companies, is based on a white fish meal, produced from very fresh flatfish. The formula food, packed in 20 kg sacks, consists for about 70% of this fish meal and for

TABLE I

Parasites and diseases in eel farming

Disease	Season	Type of eel affected	Causative agency	Control
Red fin disease:				
Aeromonas punctata	All year 28° C	Eels and elvers	Bacterial	Sulfa drugs, antibiotics, chloroamphenol 20—50 p.p.m.
Paracolobactrum anguillimortifer	All year 30.8° C	Eels and elvers	Bacterial	Sulfa drugs, antibiotics, chloroamphenol 20—50 p.p.m.
Gill disease	Summer	Eels	Bacterial	Furan or salt 0.5%
Anchorworm	All year 14—32° C	Eels (>100 g)	Parasite	Trichlorofon
Plistophorasis	All year	Elvers (1—20 g)	Parasite	?
Myxidimymiosis	All year	Eels	Parasite	Furan
Fish mold disease Saprolegnia parasitica	Spring 15° C	Eels and elvers	Fungus	Sulfa drugs, malachite green 0.2 p.p.m.
White spot disease	Spring	Elvers (0.4—5 g)	Parasite	Water sterilization by a temperature of 30° C or use of kinine
Eel dropsy disease	Spring	Eel and elvers	Virus	Vitamin E
Gas disease	Spring	Elvers (<10 g)	Plankton	Add fresh water
Trichlolinasis	Summer	Eel and elvers	Parasite	Formalin
Skin fluke	Summer	Eels and elvers	Parasite	Formalin
Tail rot	Summer	Elvers	Bacterial	Sulfa drugs
Branchio-nephritis	Winter	Elvers	Virus, bacterial or deficiency	None *
Argulus	All year	Eels	Crustacean parasite	Dipterex 0.2—0.4 p.p.m. repeatedly

* Very severe in 1970.

some 28% of so-called Alpha-starch, a partly hydrolysed product. This serves to render it possible to make a dough with cold water which does not fall apart when laid out in lumps in trays in the water of the eel ponds, not even when the eels feed vigorously and crawl into it. As additions, the formula feed contains vitamins (the so-called Halver-mixture, used in trout farming), and as minerals, the well-known Mac-Cullem mixture.

In making the dough one adds 110 parts of water to 100 parts of dry meal and adds 5—10% of fish oil, e.g., pilchard oil or cod-liver oil. For the youngest phase one uses a slightly different mixture, the main difference being the amount of vitamins admixed.

G.2. Parasites and diseases

See Table I (p. 148).

H. BIBLIOGRAPHY

Anonymous, 1970 Studies on the disease of eel culture. Shizuoka Prefectural Fisheries Experimental Station Annual Report for 1969, Hamanako Branch, pp. 265—266 (in Japanese).

Deelder, C.L., 1970. Synopsis of biological data on the eel, *Anguilla anguilla* (Linnaeus) 1758. FAO Fish. Synop., No. 80, 82 pp.

Folsom, W.B., 1973. Japan's eel fishery. Mar. Fish. Rev., U.S. Dep. Commer. 35: 41—45.

Hoshina, T. and Ookubo, M., 1956. On a fungi disease of eel. Tokyo Univ. Fish., 42: 1—13.

Inaba, D. and Uehara, R., 1962. On the types and roles played by the resting pool of eel-pond. Bull. Jpn. Soc. Sci. Fish., 28: 679—685.

Koops, H., 1966. Feeding of eels (*Anguilla anguilla*) in ponds. FAO World Symposium on Warm Water Pond Fish Culture, F.R. III-VII/E.3.

Koops, H., 1966 Die Aalproduktion in Japan. Arch. Fischereiwiss., 17: 44—50.

Korringa, P., 1970 Aalkwekerij in Japan. Visserij, 23: 315—324.

Matsui, I., 1952. Studies on the morphology, ecology and pond culture of the Japanese eel (*Anguilla japonica* Temminck and Schlegel). J. Shimonoseki Coll. Fish., 2(2): 1—245.

Matsui, I., 1964 A guide to pond culture of the eel. Marine Culture Promotion, Series 4. Japan Marine Resources Conservation Association, Tokyo, 113 pp. (in Japanese).

Ogami, H., 1971. Eel Diseases. Fish Culture Techniques, Series 1. Yogyo Times Co., Tokyo, 109 pp. (in Japanese).

Ogami, H. and Hamanaka, H., 1973. Handbook on Warmwater Eel Culture. Fish Culture Techniques, Series 4. Yogyo Times Co., Tokyo, 112 pp. (in Japanese).

Sanders, M.J., 1971. Australian studies Japanese Fish Culture Techniques. Austr. Fish., October 1971, pp. 6—7.

Takai, M. and Iohii, S., 1971. Favorite effect of artificial diets for eel. Bull. Tokai Reg. Fish. Res. Lab., 65: 53—58.

Tesch, F.W., 1973. Der Aal. Biologie und Fischerei. Paul Parey, Hamburg and Berlin, 306 pp.

Chapter 6

FARMING THE YELLOW-TAIL (*SERIOLA QUINQUERADIATA*) IN JAPAN

Example
Fish farm of Mr. Yoshiteru Matsune, Kita-Nada, Tsuchimachi, Kita Uwagun, Ehime Prefecture, Shikoku.
Period of observation: 1973—1974.
Informants: Mr. M. Wada and Mr. H. Mitsuda of the Ehime Prefectural Marine Products Experimental Station at Uwajima.

A. BACKGROUND

A.1. General principles

For the inhabitants of the Japanese archipelago, fish and other marine products form a major constituent of the daily diet: it is estimated that marine products provide 53% of the protein intake of the people of this country. The Japanese fishing industry had to expand its fishing area to meet the demands of the home market. But fish is not only considered as an indispensable source of protein, but in our time of increasing prosperity, also as a high grade food, a luxury item for the epicurean. Therefore tasty species are held in high esteem and the fishing industry often fails to find sufficient supplies to meet the demand. A high market price ensues, and this in its turn evokes the interest of the fish farmer. Farming fish can only be an economically interesting project when the market price is high, when the species under consideration is easy to handle, and grows fast on rather cheap food, and when a good supply of fry to start with is always available. In Japan, four species meet these demands so well, that flourishing farms have sprung up: the oyster, the eel, the Japanese shrimp and the yellow-tail. It is true that more species of fish are farmed in Japan, both on an experimental and on a commercial scale, but the yellow-tail, *Seriola quinqueradiata*, is by far the most important. Efforts to grow and fatten the yellow-tail in embanked ponds, starting with fry caught in nature, date back to the year 1928. But large-scale operations were only seen in the period following 1963, and now a high percentage of the yellow-tail found on the market comes from fish farms.

The year 1963 was a crucial year, because it was then that the Japanese Fisheries Resource Conservation Association, a corporate juridical body, was set up, with a membership comprising de-centralised agencies such as private fishermen's associations, commercial firms and local (Prefectural) Government groups. Its finances come jointly from membership fees and from Government grants. It bridges the gap between scientists and producers, educates the fishermen in modern cultivation techniques, and helps the Japanese Government to develop realistic fisheries policies. Of special concern was the development in the Seto Inland Seas, which maintained its high general productivity level but showed a marked decline in the catches of the more valuable species. It was the Saibai Gyogyo Centre which co-ordinated the efforts of several prefectural stations in this area and took the initiative for rearing and releasing several non-migratory species of fish and shellfish to improve their performance in this semi-enclosed area. The argument used to justify mass-release of artificially reared juveniles is that the former ecological niches are now under-exploited due to lack of natural fry. This development, of which the practical outcome is still highly debatable, was a powerful impetus for rearing some species to marketable size, and an encouraging Government policy made it possible for those interested to make a living from fish farming in management units of modest dimensions. Some 1 500 commercial units are now engaged in yellow-tail production. While the total landings of yellow-tail from the free fishery fluctuated around 40 000 metric tons in the last decade, the quantities produced by farming show a dramatic increase: 1958, 300 metric tons; 1961, 2 620 metric tons; 1968, 30 000 metric tons; 1971, 52 000 metric tons, and there are reasons to assume that these official figures are an underestimate. First it was the farms on the coasts of the Seto Inland Seas which were the leading producers, but nowadays their share amounts to only 40%, whereas the remainder is farmed on several sections of the Pacific coast in the Prefectures Mie, Ehime and Kochi, where higher water temperatures make hibernation possible.

Yellow-tail farming is a case of semi-culture starting with fry caught in nature. Efforts to rear the fry from the eggs have not yet led to practical results. One can distinguish three phases: catching the fry, rearing the fry to fingerlings of 5—10 cm length, and producing marketable fish from those fingerlings. These three phases may be carried through by one and the same company, especially if the headquarters are situated in an area where yellow-tail fry can be caught, but in other cases the fry fishers sell their catches to companies specializing in rearing fry, which in their turn hand over the fingerlings to management units producing the marketable fish.

Yellow-tail farming began in embanked ponds, which had the disadvantage of restricted water circulation, and could therefore accomodate limited numbers of fish only. This system is almost obsolete now; prohibitive construction costs being also a factor. Netting enclosures, sometimes cutting off entire bays, were the next step. Water movement was a difficult problem here as well, especially when the enclosure had to be subdivided into various compartments. Moreover, maintenance costs are high, since the nets foul easily and are then vulnerable when strong winds blow. Divers should be employed to control, clean and repair the netting regularly. Nevertheless, there are still about 70 of such enclosures in use for yellow-tail farming in Japan. The system now universally employed is based on floating netting cages. Anchored frames made of steel piping or of bamboo, supported by floats, carry bags of nylon netting in which the fish are kept. Where shelter is insufficient, anchored net cages are used, placed several metres below the surface, unaffected by waves.

That yellow-tail farming is so successful ensues not only from the high market price of the fish, but also from its extremely rapid growth: in one growth season it reaches a weight of over 1 kg — readily accepted on the market. Where hibernation is possible, this offers an economic advantage, since the price per kg goes up with the size of the fish. Contrary to expectations, the farmed fish brings a better price than yellow-tail caught in nature, because of its higher fat content and better flavour.

A.2. Biology of the species

The yellow-tail, *Seriola quinqueradiata* Temminck and Schlegel, locally called "hamachi", or "buri" when it concerns big fishes, is a torpedo-shaped fish, looking somewhat like a small tuna, but in reality belonging to the *Carangidae*, the horse-mackerels. They bear two free spines in front of the anal fin. This anal fin is shorter than the weak dorsal fin, whereas the short spines of the first dorsal fin are not free, but interconnected by a membrane. Yellow-tails are wide-ranging, fast swimming pelagic fish of carnivorous nature, feeding by day on smaller fish such as mackerel, horse-mackerel, sardines and on squids. It might be thought that such a fish would not be amenable for rearing in close confinement, but practical experience has taught that this does not present any problem. The yellow-tail is a migrating fish of subtropical nature, travelling north in spring. It is seen in the area of Okinawa in March, and travels as far north as Hokkaido. The larger specimens arrive earlier in the season than the smaller ones. The gonads of mature female fishes weighing 7 kg and over take up 10% of the body weight. More than a million eggs may be contained in such ovaries. Spawning takes place in offshore areas at water temperatures of 19—20° C. In March—April, spawning is observed off the west coast of Kyushu, and in May in Hyuga Nada, off Miyazaki Prefecture. Most of the spawning is said to take place in April and May off southern Kyushu. The eggs are small, 1.18—1.34 mm in diameter, and hatch after about 50 h. The larvae and juvenile fish are carried to inshore areas on the Kurushio current and can be found in fair numbers among floating *Sargassum* weed in various places along the Pacific coast of Japan. The yellow-tails grow rapidly at water temperatures from 20—30° C, but do not grow at all at a temperature as low as 15° C, and die when the temperature drops below 10° C. Towards the end of their first year they have reached a size of 30—40 cm and weigh over 1 kg, at the end of their second year they measure 60 to 80 cm and weigh about 4.5 kg, and they reach a length of some 90 cm at the end of a third growing season.

The fish do not feed at night, and in the dense populations in the netting cages, the conversion rate is better than when they are grown more sparsely in netting enclosures. They evidently burn up fewer calories when they have no need to move about much in search of food. They normally live in high salinity sea water, but can be grown without any trouble in water of 20—25‰ salinity. The salinity should, however, not drop under 16‰.

To protect the consumer, it should be taken into consideration that even if the farm is situated far from any industrial or agricultural pollution, pesticide residues may come in via the food chain, through food fish caught elsewhere.

A.3. Geographical situation

The farm of Mr. Yoshiteru Matsune is in Ehime Prefecture, on the southwestern coast of the Isle of Shikoku, facing Bungo Strait, which separates the isles of Kyushu and Shikoku. The headquarters are to be found some 12 km southwest of the town called Uwajima, at the exact geographical position 33°08′24″N and 132°28′18″E. The floating cages are moored in the well sheltered bay called Iwamatu Byoti, about 1 km south of the headquarters at Kita-Nada.

A.4. Hydrographic pattern

Tides. The tidal range is somewhat greater on the coasts of Bungo Strait, in open connection with the Pacific Ocean, than in many a site in the semi-enclosed Seto Inland Sea. In Iwamatu Byoti, the tidal range is 1.60 m at spring tide and 0.90 m at neaps. This is a considerable advantage, since the tides should ensure a good flushing, taking away stale water with food left-overs and droppings, and bringing fresh sea water rich in oxygen. Since the cages have netting at all sides, and are kept off the bottom, flushing is optimal.

Water temperatures. Though water temperatures do not rise so high here in summer as they do in the Seto Inland Sea, winter temperatures rarely drop to the 15°C level, so that the life of the yellow-tail is never endangered; this offers the considerable advantage that the fish can hibernate, and can be brought to the market at any time of the year, when the prices are good, whereas the farmers on the Seto Inland Sea should either market all their fish as yearlings in November or December, or should transfer them later in the year to netting cages moored further south.

The temperatures at the site of farming range from 16°C in mid-winter (February—March), to 18°C in April, 20°C in May, 22°C in June, 26°C in

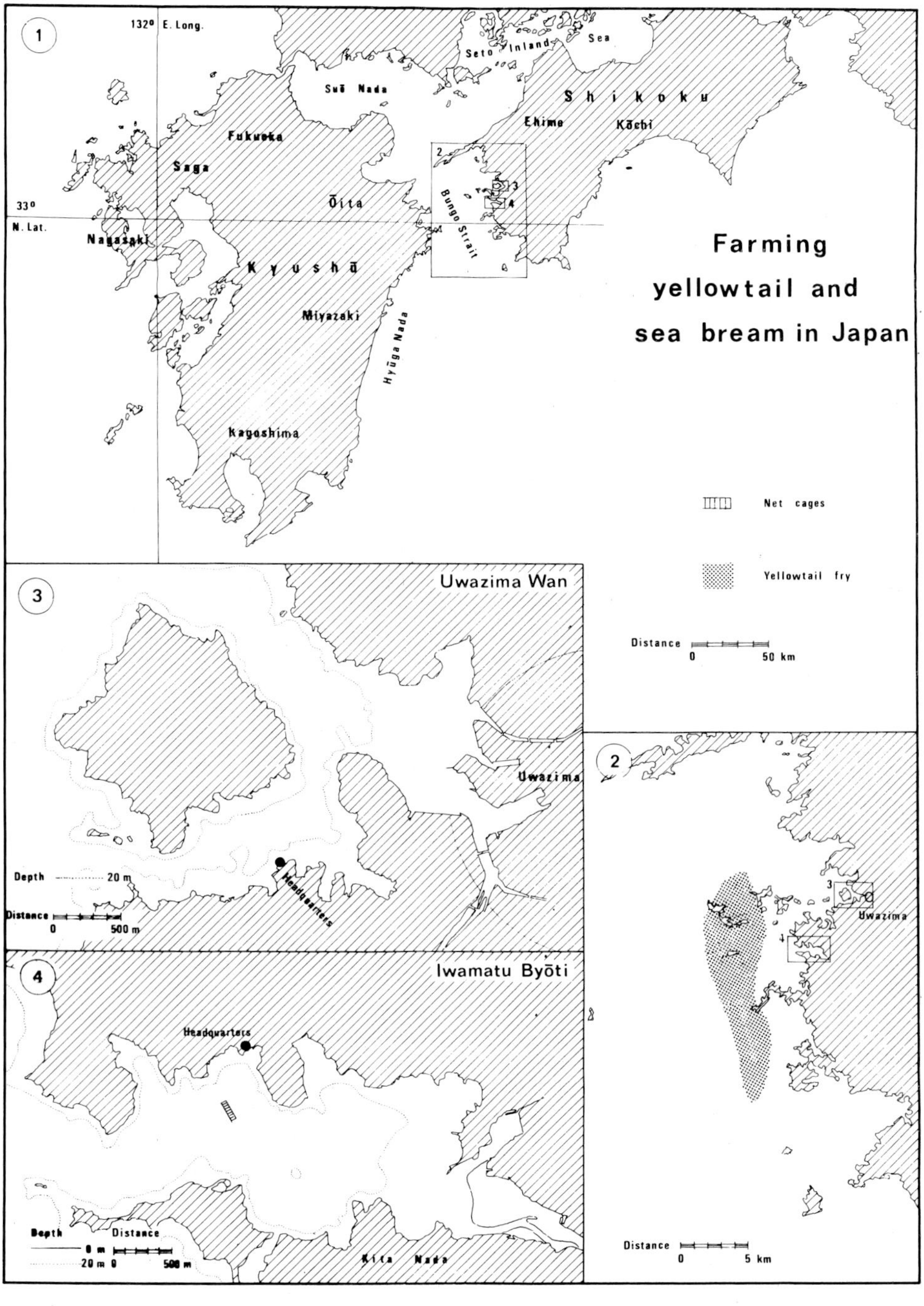

Farming
yellowtail and
sea bream in Japan
1
132° E. Long.
33°
N. Lat.
Seto Inland Sea
Suō Nada
Shikoku
Ehime
Kōchi
Fukuoka
Saga
Ōita
Bungo Strait
Nagasaki
Kyushū
Miyazaki
Hyūga Nada
Kagoshima
Net cages
Yellowtail fry
Distance 0 50 km
3
Uwazima Wan
Uwazima
Headquarters
Depth 20 m
Distance 0 500 m
2
Uwazima
Distance 0 5 km
4
Iwamatu Byōti
Headquarters
Depth 0 m 20 m
Distance 0 500 m
Kita Nada

July, 28°C in August, dropping again to 25°C in September, 23°C in October, 20°C in November and 17°C in December—January.

Salinity. The salinity is high, and very constant in this area with an average of about 33‰.

A.5. Legal aspects

Mr. Matsune's enterprise is a private business. As member of the local co-operative, he benefits from joint purchase of feed and equipment, of the joint use of some installations, including a deep-freeze storage, and of collective marketing. At times of heavy work load he can also draw on the co-operative labour pool. Mr. Matsune is chairman of the local co-operative, which has 360 member fishermen, 60 of whom farm yellow-tail and sea-bream, and 40 of whom are engaged in pearl culture.

The Prefectural Government strictly regulates the number and size of the farming enterprises. Mr. Matsune has permission to operate with 15 net cages, partly used for yellow-tail (20 000 fish), partly for sea-bream (15 000 fish).

But it is not only the actual farming procedure which is under control of the Government; capturing the fry in the open sea is also strictly regulated. The Governor of Ehime Prefecture issues 200 licences annually, with a catch limit of 30 000 per boat. The catch limit, adjusted to the national catch limit, has been set to avoid unlimited fishing for yellow-tail fry, not only to protect the natural stock from over-exploitation but also, perhaps mainly, to avoid a drop in the prices of the marketable fish by too steep an increase in the production.

B. SEQUENCE OF OPERATIONS

B.1. Catching the fry

Yellow-tail fry should be caught in nature in the appropriate season. They can be found off the west coast of the Isle of Shikoku in the area called Honda-wara, south of Uwajima, living by preference in floating fields of the gulf-weeds *Sargassum fulvellum* and *Sargassum enerve*. The official fry catching season lasts one month here, from April 20th to May 19th.

Fry fishing is carried out for 3—4 h, always during the day-time. The proximity of the Honda-wara grounds makes it possible for the yellow-tail farmers around Iwamatu Byoti to stay out for a day at a time.

Use is made of a well-boat of less than 5 tons, measuring 15—20 m in length, made of wood or of fibre-glass reinforced plastic (F.R.P.). The wells, usually a group of six, are built in the middle section of the boat, and occupy about one-third of its length. The holes in the wells are provided with fine-meshed metal gauze and every section is sheathed with a bag of fine-meshed nylon netting. The fry boats are equipped with a diesel engine of 25 to 35 h.p. Different nets may be used to catch the yellow-tail fry. Normally a fine-meshed (90 meshes/50 cm) seine net is used, 25—30 m long and 5 m deep, but dip nets with square openings measuring 1.60 × 1.60 m are also used. The fry boats carry a crew of two or three.

The catches made per day vary widely from a few hundreds to tens of thousands. The yellow-tail fry caught in the gulf-weed fields measures 2—10 cm in length, at an individual weight of a few grams only.

B.2. Transferring the fry to crawl nets

It takes 3—4 h to sail from the catching site to the farm of Mr. Matsune at Kita-Nada. The catch is transferred on arrival to crawl nets. A crawl net consists of a group of cages measuring 2 × 2 × 2 m made of fine-meshed nylon netting (90 meshes/50 cm), hung in one of the regular square frames used for farming yellow-tail described below. One of the first things that has to be done is to grade the juvenile fishes into the categories "small", "medium" and "large" to avoid losses by cannibalism. For grading one makes use of a set of rectangular wooden boxes measuring 1 × 0.50 m, with a depth of 30 to 40 cm, that are provided with a slatted bottom made of round plastic bars placed at given distances from each other. Boxes with different slit sizes are placed on top of each other in a fine-meshed netting cage. The fish fry is scooped from the crawl net into the immersed sorting boxes, care being taken that the fish are never touched by hand. Each size category of fry is placed in a separate cage.

Catch, transfer and grading, if properly carried out, does not lead to appreciable losses among the young fishes.

B.3. Rearing the fry

In the beginning the fry is fed with finely minced fresh sand-eels (*Ammodytes personatus*). The daily dose of feed is generally 20—30% of the body weight of the juvenile fish, administered during the day-time in two or three batches. Since fish with a full stomach will not come to the surface to feed, it can easily be seen if the quantities of food administered are sufficient. Some fry farmers place a light over the cage at night to attract zooplankton from the surrounding water as an extra source of food.

The yellow-tail fry grows rapidly, but the individual growth rates may differ widely, which makes it necessary to repeat the grading procedure from time to time to avoid losses by cannibalism. To that end, the whole net can be lifted up to facilitate scooping the fish into the set of grading boxes placed in another compartment.

Another matter of concern is fouling of the netting. The smaller the meshes the easier they become obstructed by algal growth or by settlement of other fouling organisms. Since perfect flushing of the net cage is of paramount importance the netting should be changed about every 10 days. A simple and efficient system is to fold the net cage somewhat together, to place a clean net adjacent to it, to tie them together at the top side, to lower the joint section, and to lift up the old net so that all the young fish can be poured into the new net, which can then be extended to its full size when the dirty net has been taken away. The latter must be properly cleaned and dried for repeated use.

The size of the net cages and the mesh size of the netting should be adjusted to the size of the fish. Up to 5 cm body length, net cages of 2 × 2 × 2 m with netting of 90 meshes per 50 cm * are used; for fishes from 5—10 cm length, cages measuring 3 × 3 × 3 m and netting of 18 meshes per 15 cm, which provide better flushing, are used. In 4—6 weeks the fish reach a size of 5—10 cm and a weight of 8—50 g each.

B.4. Farming fingerlings to marketable size

When the fishes have reached a size of about 10 cm, they are transferred to bigger cages with still wider meshes: 5 × 5 × 5 m and 10 meshes per 15 cm. Many farmers buy the fingerlings at this stage from enterprises specializing in rearing of the fry. The fingerlings can be transported over considerable distance in the same well-boat used for catching the fry. The company under consideration does not buy fingerlings, but performs all three operations: catching the fry, farming fingerlings and rearing fingerlings to marketable size.

The basic equipment is the floating net cage. Such a unit consists of a square frame measuring 10 × 10 m, made either of galvanized steel tubes or of bamboo. This frame is kept afloat with styrofoam buoys, sheathed in polyethylene bags to prevent fouling and placed under the corners, or by 40 gallon drums. A cat-walk of planks is installed on top of the frame to facilitate the work. A net cage of knotted nylon, measuring 10 × 10 m, 8 m deep, is hung in such a framework and weights are fastened to the bottom to keep the cage fully stretched. Usually a series of these units is lined up and fastened together, duly anchored in a suitable place, keeping the cages well off the bottom.

The frames usually measure 10 × 10 m, accommodating either one big net cage or a group of smaller ones. For fishes of 10—25 cm in length, one uses cages measuring 5 × 5 × 5 m with netting of 10 meshes per 15 cm, for fish of 25 cm and over, cages measuring 10 × 10 × 8 m with netting of 6 meshes

* For fish up to 5 cm one uses netting of 90 "kei", which means 90 square meshes per 50 cm. For the larger fish the netting has lozenge-shaped meshes, for which the "fushi" unit is used: the number of meshes per 15 cm.

per 15 cm. Interchanging of fouled nets remains a necessity. The number of fishes per m^3 of cage depends on their size, as indicated in the following table:

Weight of the individual fish (g)	Number of fishes/m^3
<25	115—340
25—100	58
100—200	45
200—400	23
400—600	14
600—1 000	10
>1 000	6—7

The fishes are fed in all phases with minced fish, usually kept in storage in deep-frozen form. It is of importance to select the right type of fish, avoiding as far as possible fat fish such as anchovy. Too much anchovy unavoidably leads to metabolic disorder and death. Fish species widely used as food

Fig. 65. General view of the netting cages for fish farming at Kita Nada. (Photo P. Businger.)

Fig. 66. Frozen blocks of feed fish, mainly saury, to be used on the yellow-tail farm. (Photo Y. Akimoto.)

Fig. 67. Frozen and defrosted saury for feeding yellow-tail. (Photo Y. Akimoto.)

in the yellow-tail farms are sand eel (*Ammodytes personatus*), saury (*Cololabis saira*), and mackerel (*Scomber japonicus*). Horse-mackerel (*Trachurus japonicus*) and sardines are also used, and, because of its low price, also some anchovy. The fish is usually bought as deep-frozen blocks, kept in storage at —25°C, in a section of the freezing house of the local fisheries cooperative, for which rent has to be paid.

The frozen fish has to be thawed out, and is then minced mechanically. The minced fish is led into a shallow boat to be transferred to the net cages. Using a long-handled shovel, a worker administers the food to the cages, scattering it thoroughly (Fig. 69). Larger fish require about 10% of their body weight per day in food, administered in one, sometimes two, batches. The yellow-tail are fed every day, Sundays included, but in the colder winter months they are fed only every two or three days.

The graph in Figure 70 indicates the daily feed requirement in percentages of the body weight and shows the increase in body weight in the course of the first summer season. Growth is fast and the conversion rate is said to fluctuate between 7 and 8, much better than in shrimp farming, but poorer than in farming chickens and pigs in the bio-industry on land.

Fig. 68. The defrosted fish passes the mechanical grinder and slides down into the boat used to feed the fish in the cages. (Photo Y. Akimoto.)

Fig. 69. The yellow-tail is fed by throwing minced fish into the net cages with a shovel. (Photo P. Businger.)

Survival is remarkably good in yellow-tail farming. If the farm is properly managed one may count on 70—80% survival for the whole cycle. One man can manage from 10—15 units, considered as a family size farm.

B.5. Use of other nets

When the farmer is fortunate enough to be able to grow the yellow-tails in an almost perfectly sheltered area, floating net cages are appropriate equipment. In less sheltered areas damage by gales should be feared. A recent development is the use of submerged netting cages, by which, such damage can be avoided. Large rectangular netting cages, up to 20 m in length and 10 m in height are anchored in such a way that they are kept well above the bottom and five or more metres below the surface of the sea. Anchors and buoys keep the net cages fully extended. A square funnel is constructed in the centre of the ceiling of the net. This funnel serves for feeding the fish, which will swim up into it at feeding time. In case of rough weather, the funnel can be lowered. Waves will not easily damage such nets if properly installed. Additional advantages are that sites with a fair current, offering excellent flushing, can be selected, and that nobody can see how many fish are being reared.

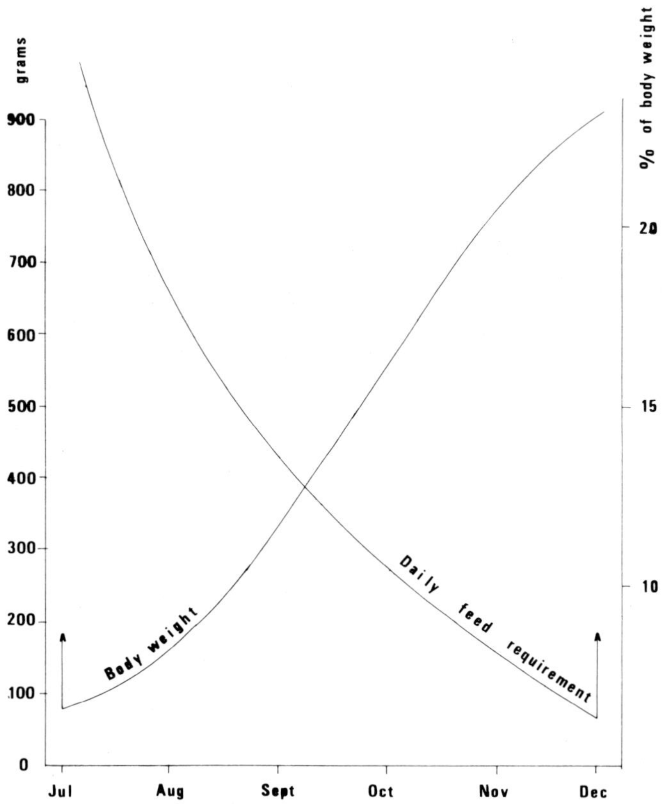

Fig. 70. Feed requirement and increase in body weight in the yellow-tail. (Courtesy Mr. H. Mitsuda of the Ehime Prefectural Marine Products Experimental Station at Uwajima.)

B.6. Harvesting and grading the yellow-tails

Towards the end of the first year of farming, the yellow-tail will have reached a weight of 0.8—1.2 kg and will bring a price of about Y 500 per kg. At that time, considerable quantities are marketed, especially from the farms in the Inland Sea area, with a noteworthy peak in November. Mr. Matsune, operating in warmer waters, keeps most of his yellow-tails for a second year, when they reach 3.0—4.0 kg, bringing a price of Y 650 per kg, whereas they reach 7.0—8.0 kg and a price of Y 800 per kg in a third growing season.

When it is decided to harvest the crop, a motor-boat, equipped with booms and power winch, is brought alongside a netting cage, and a group of about eight workers begins to pull up the net cage by hand, detaching it from the frame at the far side first, and gradually pulling it towards the boat (Fig. 71). Once the net is entirely alongside the boat, it is pulled together in one corner and lifted out of the water to the extent that the fish become

Fig. 71. The first phase in harvesting yellow-tails from a net cage: the net is detached at the far side of the frame and gradually pulled towards the vessel. Note the steel net frame in the foreground. (Photo Y. Akimoto.)

visible in great concentration. A long-shafted landing net is then used to scoop up the fish (Fig. 72). This net, with a diameter of about 80 cm and with a bag of some 100 cm deep, guided by hand, is suspended from a line leading to the winch via a boom, so that it can be lifted up mechanically when filled with fish (Fig. 73). The batches of fish scooped up from the net cage are carefully weighed and next emptied onto a sloping sorting table (Fig. 74) provided with side boards, from which any ill-shaped, smallish or damaged specimens can be quickly eliminated. The remainder slides from the sorting table into a shallow cage-net installed on the port side of the boat, for the fish must be kept alive.

B.7. Transfer to the market

When all the fish to be sold are gathered in shallow crawl nets, a 30-ton transport vessel equipped with a partioned well (Fig. 75) comes to take them up, use being made again of the same scoop net described above. This vessel is owned and operated by a private company, not associated with Mr. Matsune's enterprise. Usually the fish are shipped by night to the market, normally the Hiroshima area. Proper timing is of the utmost importance in

Fig. 72. When the net cage is drawn together in a corner, the landing net is lowered into it. (Photo Y. Akimoto.)

Fig. 73. The scoop net, well-filled with yellow-tails, is hoisted up with the winch. (Photo P. Businger.)

Fig. 74. A batch of yellow-tails to be dropped on the sorting table. (Photo P. Businger.)

Fig. 75. The transport vessel, equipped with a partitioned well, brings the yellow-tails to the market. In the foreground is a bamboo net frame. (Photo Y. Akimoto.)

transporting and marketing yellow-tail, for the fish should be as fresh as possible when landed. Usually the fish is killed on board and put in boxes on deck about 6 h prior to marketing, but it is also common practice to sever the spinal cord with one stab of a knife which does not kill the fish, but merely paralyses it.

B.8. Consumption

The favourite dish from yellow-tail is "sashimi", consisting of thin slices of raw meat, dipped in soy sauce. Yellow-tails may, however, also be grilled or used in soups.

C. FARMING RISKS

C.1. Hydrographic conditions

In selecting a site for installing net cages, one should take into consideration that good flushing, supplying water rich in oxygen and eliminating waste, is of paramount importance. Shelter is, however, also a factor to take into account. Submerged net cages can be used to advantage where shelter is insufficient.

C.2. Predators

Since the yellow-tails are farmed in monoculture in net cages, predation does not occur. In areas where sharks abound, the nets may be torn open. It is then wise to use either netting made of metal wire or to protect the whole farming area by huge fences of wide-meshed wire netting.

C.3. Parasites and diseases

Several parasites are known to attack yellow-tails in the net-cages, especially when the cages are overcrowded or water circulation is poor.

1. *Benedinia seriolae*, a flatworm attacking the skin of the fishes. If numerous, it may weaken the fish and cause loss of appetite. These parasites can be controlled by bathing the affected fish once a fortnight in fresh water, for 3 min at a water temperature of 26°C, for 5 min at 16°C.
2. *Heteraxine heterocerca*, a gill parasite attacking the fishes at water temperatures of 24°C and over, to be controlled by dipping the fishes for a few minutes in saturated brine.
3. *Caligulus amphifurens*, a parasite of the oral cavity.
4. *Brachyura seriolae*, a parasite living on the fins, leading to skin discolouration.

5. A Metacercoid bladderworm, reaching the yellow-tails via fresh anchovies used as feed.

6. *Ichthyosporidium*, a fungus which may affect the circulation system and some other organs of the fish.

7. *Vibrio* species causing vibriosis, a bacterial disease usually considered as secondary infection, to be cured with sulfa drugs. Vibriosis hardly occurs under perfect feeding conditions.

In addition to the impact of these parasites, diseases to be considered as metabolic disorders are known. Feeding too much anchovy does lead to such disorders and ultimately to death, presumably because the yellow-tails cannot properly cope with an excess of unsaturated fatty acids. In the juvenile specimens one may observe urinary stones as a consequence of nutrient deficiency.

D. ECONOMIC ASPECTS

D.1. Rents

As a member of the local co-operative fisheries association, Mr. Matsune has to pay an annual fee for use of the site where he moors his floating cages, but this is not a heavy burden on his budget. As usual the fees of the co-operatives are adjusted to the size and the proceeds of the enterprise.

D.2. Inventory items

Of considerable importance is the well-boat used to fish the yellow-tail fry. This 5-ton boat, equipped with a diesel engine, cost about Y 5 000 000 and has a life estimated at only 5 years, whereas a boat of reinforced plastic would last much longer. This boat is only used for part of the year. If Mr. Matsune did not himself fish the yellow-tail fry, he would be obliged to buy such fry for between Y 50 and Y 100 for fry of 5 cm length.

For fishing the fry, a fine-meshed purse seine and some scoop nets are also needed.

For the floating cages, material is needed to construct the frames. Metal frames last longer than bamboo frames, but cost more. Mr. Matsune operates with 15 floating frames, partly used for yellow-tail farming, partly for rearing red sea-bream. In the frames, netting cages are installed, either single or partitioned, made of netting of different mesh size. Since the nets need to be cleaned repeatedly, there should be a good number of them on hand. Fine-meshed nylon netting is more expensive than that with coarser meshes, and moreover, lasts a shorter number of years: coarse nets have a life of 5 years, fine nets of approximately 3 years. Some 18—20% of the annual expenditure should be allotted to material for the floating cages. A net cage of 2 × 2 × 2 m costs about Y 18 000; one of 3 × 3 × 3 m, Y 31 000; one of 5 × 5 × 5 m, Y 90 000; and one of 10 × 10 × 8 m, Y 280 000. The largest netting cages, measuring 16 × 16 × 9 m, cost about Y 500 000. The price per square metre of nylon netting varies according to the mesh size from Y 600 to Y 900. Through collective purchase of equipment, Mr. Matsune's enterprise is sure to pay reasonable prices.

D.3. Expendable items

Of the expendable items, feed is by far the most important. On the average, one may assume that some 50% of the annual expense goes to purchase of feed, although the percentage is even higher if the enterprise is larger. For every kg of yellow-tail produced, one should count on 7—8 kg of fish to feed them. In 1971 and 1972 feed cost Y 30—Y 50 per kg, depending on the type of fish, the supply, transportation costs and whether it was fresh or deep-frozen. In 1972 there was a good supply of frozen saury from Hokkaido and northern Honshu. Anchovy is usually quite cheap but leads to metabolic disorder when used on too large a scale. Recently the price for feed went up steeply, as a consequence of high fish meal prices all over the world, this being principally due to a serious drop in the anchoveta catches in Peru and to high fuel prices. Considering the great part feed constitutes in the budget, such a development may easily endanger the remunerativeness of yellow-tail farming.

All fish farming becomes a hazardous proposition when the price for feed approaches 10% of that to be made for the marketable fish.

The co-operatives organize collective purchase of feed for their members, and since the State helps in establishing such facilitates, can provide deep-freeze storage at a very reasonable cost.

The quantities of fuel for the engines used by the yellow-tail farms are modest.

D.4. Personnel and wages

Mr. Matsune works with a permanent staff of two, one male and one female, but at time of heavy work load, for instance in the harvesting season, he can request assistance from the labour pool of the local fisheries co-operative. The costs of labour are of the same order of magnitude as given for other cases of farming the sea in southern Japan. Wages weigh far less heavily on the budget than does feed.

D.5. Sales

The price on the market for yellow-tail fluctuates rather widely with the season and with the size of the fish, viz. between Y 400 and 800 per kg.

Cultivated yellow-tail, now 60—70% of the market, or according to Furukawa, even 90% of the "hamachi" (the smaller specimens), makes a price 10—20% higher than stock caught in the wild, simply because farmed yellow-tail is fatter and has a better flavour. Yellow-tails farmed during one season only and weighing 0.8—1.2 kg each, make Y 500 per kg, those of 2 years, weighing 3—4 kg each, Y 650/kg, whereas 3-year-old yellow-tails weigh 7—8 kg each and bring Y 800 per kg. To calculate the price the producer gets, one should subtract 5% for marketing costs and 2% for transportation. Mr. Matsune sold in 1973—1974 some 30 000 kg of yellow-tail per year, a gross income of Y 18 500 000. Considering that every kg of yellow-tail requires 7—8 kg of feed fish, at Y 40 per kg, one can easily see why about half of the budget is required for purchase of feed. Since a production of 7 000 to 10 000 fish is considered as a reasonable management scale, it can be concluded that Mr. Matsune's enterprise is a fair-sized one, especially when one takes into account that he farmed, at the time of observation, not only some 20 000 yellow-tails, usually grown for 2 years, but also some 15 000 sea-breams, which bring a very good price, but have to be farmed for 3 seasons.

The production of farmed yellow-tail in Japan was officially estimated at some 30 000 tons for 1968, but insiders claim that the real tonnage may now amount to about triple that value.

E. GOVERNMENT ASSISTANCE

Without the positive encouragement from Governmental Agencies, yellow-tail farming would surely not have reached the present high level of production. It is difficult to analyze exactly how much of the effort should be ascribed to the Government, since it consists not only of the scientific research provided free of charge to assist the farmers in problems of a biological and technical nature, but also the general policy to subsidize the fisheries co-operatives, and to compel the farmers to be members of such co-operatives so that they can profit from collective purchase of material and feed, and from collective marketing, which protects them against speculation and against too fierce competition. Such co-operatives can build large units for freezing and deep-freeze storage much more efficiently than could the individual farmers, the more so as the Government subsidizes construction and purchase of this equipment quite heavily.

The policy of the Government made it possible that some 1 500 commercially sound units engaged in yellow-tail farming came into being in a short time, creating a living for many people and providing the market with a fair tonnage of a much sought-after item.

It is also of importance to note that the Government considered it necessary to regulate catching of yellow-tail fry in the open sea; explained as a measure to avoid over-exploitation of the natural stock, but perhaps rather to aim at a gradual growth of the industry.

The yellow-tail farmers of Ehime receive valuable assistance from the Ehime Prefectural Experimental Station at Uwajima.

F. GENERAL IMPRESSION

Everyone who has made acquaintance with yellow-tail farming as carried out in Japan cannot help being impressed by this new industry which sprang up so rapidly and now provides the market with an impressive tonnage of a much sought-after product, compared with which, the catches made of the same species in the wild almost dwindle to insignificance. The extremely rapid growth in this species and the high water temperatures prevailing in the waters of southern Japan are the basis of this success. This surely is, for Japan, an interesting development, both to the benefit of the discriminating fish consumer and of those who now make a good living this way. Considering the great number of fish kept in the netting cages, Bardach et al. (1972) even exclaimed that "the yield of yellow-tail must be among the highest ever obtained through aquaculture". But is this a sound approach? Should one not compare yellow-tail farming with a modern bio-industry on land, producing poultry, veal or pork? In such cases one would certainly not boast of production figures obtained per hectare, since the feed, which is only converted here, comes from a much greater acreage. The same holds good for yellow-tail farming; one may rear a great number of those in a relatively small netting cage, but the feed they thrive on comes from a much vaster volume of water.

It is even doubtful whether one should say that the Japanese yellow-tail farms contribute positively to supplying the population of the Japanese archipelago with valuable food, rich in protein, and that every effort should be made to encourage developing countries, short in animal protein, to follow that example. The point is, that yellow-tail farming does not really produce animal protein, but rather, upgrades animal protein by converting low grade fish into high grade fish. One should take into consideration that the conversion factor is 7 or 8, which means that one needs 7 or 8 metric tons of low grade fish to produce 1 metric ton of yellow-tails. This is perfectly all right when a rich country, such as Japan, can afford it, and wishes to pay a high price for a luxury product, but if the time ever came that it was difficult to provide the population of the Japanese archipelago with the required minimum of animal protein, the Government would have to take the hard and unpopular decision to forbid farming of yellow-tails, sea-breams, eels and shrimps. The feed given now to the yellow-tails predominantly consists of species such as sauries, horse-mackerels, mackerels, sardines and sand eels, which are perfectly edible for man, if treated properly aboard and ashore, but which are considered as less palatable by the discriminating consumer. This is the reason that one should never talk developing countries, short in animal protein, into following the Japanese example, unless they could find such a good market abroad that they could buy a multiple tonnage of low grade but wholesome animal protein in return.

Fortunately, Japan can still afford to encourage its farmers to produce high grade fish at the cost of a much larger tonnage of low grade fish, and yellow-tail farming may continue to flourish if the consumers are willing to pay a higher price than they used to, now that the price for feed fish has gone up so much through international developments in fuel and fish meal prices. There are two constraints in yellow-tail farming: potential shortages in feed and in fry. The quantities of feed required for the expanding yellow-tail industry are great, and it might become difficult to purchase the quantities needed at a reasonable price. Therefore, research is already under way to switch over from fresh fish to pellets, in which soy meal and yeast could perhaps be incorporated; however, yellow-tails require a food containing at least 70% of protein, and thus far the results obtained with pellets fall far behind those with fresh fish as food. It is certainly not impossible, however, that a suitable pellet can be developed for yellow-tails. Nevertheless, such pellets will never be quite cheap, since fish meal will most probably be an indispensable ingredient.

The quantity of fry to be found in nature may fluctuate from year to year and from place to place, but nature seems to still allow an annual catch of several tens of millions of juveniles to be transferred to the yellow-tail farms. The Japanese Government has regulated the catch of fry, but it seems doubtful whether there exists such a clever specialist in the field of population-dynamics that he can forecast the number of juveniles one can catch safely without interfering with the natural reproduction of this species. It seems more likely that the primary objective is to prevent so rapid a growth of this industry that market prices would drop to a level which would make remunerative farming impossible.

The farmers feel that shortage of fry might interfere with their efforts, and they urge the Government to spend more effort in the field of artificial reproduction of yellow-tails. There is no reason why it should be impossible to rear fry from the eggs, but so far great difficulties are being experienced because this species has small eggs, which means that extra care has to be taken in providing the right type of food for the earliest phases.

The whole problem of rearing fish fry requires more scientific attention, especially when one plans to release juvenile fish in a natural environment. Doubt has been expressed from several sides (e.g. by the Japanese expert, Fujiya) whether such fry, which never had to search for food or to hide or flee from predators, would be so pampered by prolonged rearing under artificial conditions that they would never be able to survive under natural conditions. Experience with release of salmon smolts, and the consideration that few species of fish ever have the privilege of being instructed by their parents seems to point, however, in another direction.

The yellow-tail industry, yielding a high quality food to those who can afford it, and offering a livelihood to a great number of farmers operating with small management units, could certainly profit by more research on the rearing of the fry, the development of pellet food, and control of parasites and diseases.

G. BIBLIOGRAPHY

Bardach, E., Ryther, J.H. and McLarney, W.O., 1972. Aquaculture. The Farming and Husbandry of Freshwater and Marine Organisms. Wiley-Interscience, New York, N.Y., 868 pp.

Furukawa, A., 1971. Outline of the Japanese Aquaculture. Jpn. Fish. Res. Conserv. Assoc., Tokyo, 39 pp.

Harada, F., 1965 Studies on propagation of yellow-tail (*Seriola quinqueradiata* T. and S.) with special reference to relationship between feeding and growth of fish reared in floating crawl net. Mem. Fac. Agric. Kinki Univ. (Kinki Daigaku, Nogakubu Kiyo) 3: 1—291.

Harada, T., 1970. The present status of marine fish cultivation research in Japan. Helgol. Wiss. Meeresunters., 20: 594—601.

Kimura, K., 1952. Study on the spawning ground and season of yellow-tail, *Seriola quinqueradiata* Temminck and Schlegel, determined by ovary weight. Bull. Tohoku Reg. Fish. Res. Lab., 1: 54—62 (in Japanese with English summary).

Milne, P.H., 1973. Development of fish farms in Japan. Fish. News Int., 12(12): 17—24.

Minamizawa, A., 1971. Studies on crawl net for hamachi. Ehime Prefect. Mar. Prod. Exp. Stn., Yoshoku, 22 pp. (in Japanese).

Ohiwa, Y., Matsusato, T., and Saki, H., 1971. Disease and treatment of hamachi. Ann. Rep. Ehime Prefect. Mar. Prod. Exp. Stn. for 1970, pp. 81—108 (in Japanese).

Tamura, T., 1970. Marine aquaculture. Nat. Sci. Found. NSF, Washington, D.C., P.B. 194, 051T.

Chapter 7

FARMING THE RED SEA BREAM (*CHRYSOPHRYS MAJOR*) IN JAPAN

Example
Fish farm of Mr. Sadanobu Imura, Sakashizu, Uwajima, Shikoku.
Period of observation: 1973—1974.
Informants: Mr. M. Wada and Mr. H. Mitsuda of the Ehime Prefectural Marine Products Experimental Station at Uwajima.

A. BACKGROUND

A.1. General principles

Among the multitude of fish species used as human food in the Japanese archipelago the red sea bream, locally called "ma-dai", occupies a very special place. It is held in high esteem and the consumer is willing to pay well over Y 2 000 per kg for specimens of good size weighing over 1 kg. This is not so much because of its exquisite flavour or nutritional value, but rather since it is traditionally associated with "good fortune". Therefore "ma-dai" is the favourite fish for birthday celebrations and weddings. It is a kind of symbol like turkey in the United States. Usually one serves two well-sized specimens on a platter, adorned with plum blossom if the season allows it. The favourite preparation is as "sashimi": after carefully removing the skin from one flank, the raw fillet is taken from the backbone, cut into thin slices, laid back on it slice by slice, to be covered again by the skin. One dips the slices of raw meat in soy sauce to enrich its flavour. The meat of the other flank of the fish is later used for soup. Occasionally the red sea bream is prepared in other ways, e.g. by grilling.

The catches of red sea bream were quite poor up to about the year 1920, when motor-driven boats came into use for bottom trawling. This led to a marked increase in the catches, with top years from 1922 to 1926. Thereafter the catches declined dramatically, much to the regret of those who, at celebrations, wished to adorn their table with this red-coloured fish.

The high price the consumer is prepared to pay, based on the great demand for special occasions, led to efforts to rear this species artificially, starting as early as 1887. In 1902 a special hatchery for this species was set up at Onomichi, in Hiroshima Prefecture, but that did not last long. It is only from 1962 onwards that the rearing techniques have been so well established that one can grow a decent quantity from the egg. The red sea bream was one of the species it was desired to re-establish in the Seto Inland Sea. The Saibai Gyogyo Centre envisaged large-scale rearing of juveniles, to be released in the Seto Inland Sea to occupy the former ecological niches now thought of as under-exploited due to lack of natural fry. No scientific evidence has, however, been adduced that this releasing policy led to strengthening of the stock, but fish farmers took an interest in the possibility of rearing the juvenile fish produced by the hatcheries to consumption size, despite the rather slow growth of the red sea bream. Most of the juveniles now reared in the hatcheries are bought by fish farmers in the area. Still, hatchery fry occupies, thus far, only a modest place in the provision of fish farms with juvenile red sea bream. By far the largest quantity of fry is caught in the wild, to be transferred to the fish farms in well-boats. Farmed sea breams are not held in such high esteem as specimens caught in the wild, which display a more attractive red colour and are said to have a better flavour. For wild stock the price may be Y 3 000 per kg.

The total tonnage of red sea bream farmed in Japan falls far short of that of the yellow-tail. In many cases, like in the example chosen, the farmer does not specialize in sea bream only, but combines yellow-tail farming with it.

A.2. Biology of the species

The red sea bream, *Chrysophrys major* Temminck and Schlegel, is a fish belonging to the Sparidae, a family of subtropical and tropical fishes characterized by a laterally compressed body, bearing quite large ctenoid scales, one long spined dorsal fin, a spined anal fin, soft-rayed ventral fins placed under the pectoral fins and a forked tail fin. The Sparidae are equipped with powerful teeth and molars, enabling them to crush the shells of various invertebrates. Many species are held in high esteem by man because of their palatable flesh.

The red sea bream is found in the Japanese archipelago in inshore waters with gravelly or rocky bottom, predominantly at a depth of 80—90 m, migrating to shallower sites of 30—40 m depth in the spawning season in May and June. Though the red sea bream avoids the cooler waters of northern Japan, and is only rarely caught off Hokkaido, it can stand temperatures as low as 5° C.

The red sea bream is a demersal fish, feeding on a variety of bottom invertebrates which it can overpower and crush with its powerful set of teeth. It grows rather slowly and loses its appetite at temperatures lower than 14° C, to stop feeding completely when it drops under 11.5° C, leading to hibernation. The red sea bream can reach a size of over 80 cm at a weight of about 8.5 kg, but such specimens are rare and are said to be over 25 years of age.

The red sea bream spawns its pelagic eggs in May and June, at water temperatures from 17—19° C. Hatchery experience has shown that incubation will take some 54 h at 18° C but only 35 h at 22° C. Specimens under 2 cm in length are rarely observed in nature.

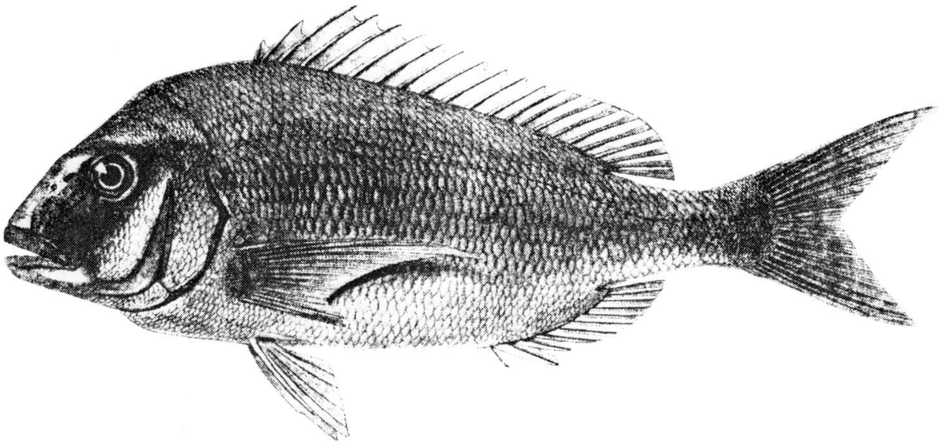

Fig. 76. The red sea bream, *Chrysophrys major.*

A.3. Geographical situation *

Mr. Imura's plant is located 3 km due west of the town of Uwajima, on a small peninsula in a well-sheltered inlet of Uwajima Bay in the northeastern part of Bungo Strait. The exact geographical position is 33°12′50″N and 132°32′05″E. Uwajima is situated in Ehime Prefecture on the western side of the isle of Shikoku in southern Japan.

A.4. Hydrographic pattern

The hydrographic pattern and the topographical situation in the section of Uwajima Bay where Mr. Imura's plant is situated differ so little from those at Kita-Nada, described in the chapter on yellow-tail farming, that detailed data need not be presented here. The tidal range is practically the same, the salinities fluctuate but little, and showed in the year 1973 an average of about 32‰. Water temperature was here, on the other hand, somewhat lower in the winter of 1972—1973: January 1973, 13.2°C; February, 12.8°C; March, 12.5°C; April, 15.3°C; May, 18.6°C; June, 21.1°C; July, 24.5°C; August, 26.6°C; September, 24.7°C; October, 22.0°C; November, 19.1°C; and December 1973, 15.4°C.

Mr. Imura's net cages are not anchored in open sea, but to a wooden pier.

A.5. Legal aspects

In many respects the legal aspects of Mr. Imura's enterprise closely resemble those of the fish farm of Mr. Matsune described in Chapter 6. Mr.

* See Chart in Chapter 6.

Imura operates on private basis with 15 rearing units moored to a wooden pier. In some of his netting cages he grows red sea breams, and in others yellow-tails. Mr. Imura's enterprise includes a freezer of 200 metric ton capacity in which he stores fish from his own farm and from others. He mostly handles his own transportation and operates with a considerably larger personnel than Mr. Matsune.

B. SEQUENCE OF OPERATIONS

B.1. Hatchery operations

Though Mr. Imura does not operate a hatchery himself, and works predominantly with fry purchased from Kyushu fishermen, it is interesting to present some details on hatchery operations for red sea bream. The hatcheries have produced in recent years some 2 000 000 juveniles, the majority to be purchased by fish farmers. Part of this red sea bream fry is produced in the Uwajima Experimental Station. Mr. M. Ukawa of the Hakatajima Experimental Station of Ehime Prefecture should be given credit for developing adequate hatchery techniques for the red sea bream.

Wild brood stock are used by preference, the male fish being recognized by a more angular shape of the head and a darker pigmentation. Since the red sea bream is a rather recalcitrant spawner in captivity, requiring intramuscular injections with hormone composites to bring it to the mood, the eggs are usually fertilized artificially on board ship, immediately after the mature adults are caught. The fertilized eggs are then transferred to the hatchery with great care, avoiding intense light, shocks and vibrations as far as feasible.

Since the eggs of the red sea bream, 500 000 of which may be produced by a 5-year-old female, are pelagic, it is of paramount importance to fill the tanks which should receive them with sea water of the correct specific gravity, not lower than 1.0245 at 15°C, to prevent the fertilized eggs sinking to the bottom. A tank of 1 or 2 m^3 capacity, receiving subdued light (less than 3 000 lux) through translucent plastic plates is used for hatching some 50 000 eggs, which takes 54 h at 18°C, but only 35 h at 22°C. The newly hatched larvae measure 2.3 mm in length and live for the first 4 days on the reserve food stored in the yolk sac. The water in the tank is, in principle, recirculated to avoid the entrance of predators with the fresh sea water, but some new water is taken up every 10 days, so that about half the volume of the tank is renewed within one month. The usual rearing temperature is about 20°C.

When the yolk sac is resorbed the larvae should be fed with items small enough to be ingested. Usually the first food given consists of trochophora larvae of *Crassostrea gigas*, which can easily be reared; next follows the small rotifer *Brachionus plicatilis*, which is 0.1 mm. *Brachionus* is, in its turn, reared in a series of 900 l tanks, in water enriched with nutrient containing a thick culture of the minute alga *Dunaliella*. A density of 70 *Brachionus* per ml may be reached; the culture is cropped with fine-meshed silk sieves to be fed to the sea bream larvae. Some four *Brachionus* tanks may be needed to produce enough feed for 100 000 sea bream larvae, for at a size of 3.5 mm, one sea bream larva may devour up to 100 *Brachionus* per day. After about the 10th day the juvenile larvae are big enough (about 5 mm) to grasp and swallow nauplii of the brine shrimp *Artemia salina*, which can easily be reared from dry-stored eggs, these being imported from the United States. *Artemia* eggs are, however, quite costly, at Y 10 000 per 2 kg, 1 g holding some 260 000 eggs. From the 20th day onwards, the sea bream larvae assume benthic habits and can be offered small polychaetes and finely minced shrimp meat as food, followed, in a later phase, by minced flesh of lean fishes. As soon as possible the juvenile fishes should be transferred to outdoor net cages; but only under very favourable current and wave conditions can one venture to do so for juveniles of a length of under 20 mm. A continuous hazard during rearing is deterioration of the water due to biodegradation of excess feed and of dead specimens.

B.2. Procuring the fry

The overwhelming majority of the sea bream fry used for farming is still collected in nature. Even though fish farmers foresee a shortage of fry in the future, they can still get a supply from the fishermen operating from ports in northern Kyushu. In the appropriate season, the fishermen search for juvenile sea breams in offshore shallow waters and try to detect them by eye or with the aid of an echosounder. The juvenile fish they catch there measure 2—3 cm in length. They are transported in their well-boats, stored in crawl nets and in due course, sold for a price of Y 130—150 each. The fish offered for sale have mostly been kept for some time in net cages and measure 2—5 cm in length. Red sea bream fry can be purchased in several ports in Fukuoka, Nagasaki, Saga and Oita Prefectures, but Mr. Imura invariably obtains his fry from dealers in Hatanoura, Oita Prefecture.

B.3. Rearing the fry to marketable size

The rearing system for red sea bream is identical with that for yellow-tail (use being made of the same net cages and the same feed fishes, shunning fat fishes such as anchovy as far as feasible) and therefore a detailed description will not be given here. The net cages used by Mr. Imura measure 8 × 8 × 8 m, the frame being constructed of wood. They are kept afloat by styrofoam buoys. Of his 15 cages, three were used for red sea bream at the time of observation and 12 for yellow-tail.

Fig. 77. The net cages of Mr. Imura's farm are moored to a wooden pier. (Photo P. Businger.)

For red sea bream the following numbers should be stocked per m^3:

Individual weight (g)	Number per m^3
25	150—300
100	45
200	23
400	14
1 000	6—7
2 000	3

Until it has reached a weight of 200 g, the sea bream is fed 20% of its body weight per day in the summer season. Beyond that stage the ratio becomes 10%, whereas in winter the ratio is kept as low as 0.5%. Throughout the whole development the same type of food and the same system of feeding described in Chapter 6 are used. During the last 10 to 20 days of farming, preceding the sale of the marketable fish, some special food can be administered to obtain the pink colour of the flesh, which is so highly appreciated by the discriminating consumer. To this end one may use deep frozen Euphausids, collected from stomachs of whales, caught in arctic or antarctic waters. This food is quite expensive.

At an age of 2 months, the sea bream may have reached a length of 20 mm, at 6 months 10—15 cm, at one year (in spring) 20 cm, at 2 years 30 cm. Sea bream fry of 15 mm, as purchased from a hatchery, should be placed in cage of netting of 260 meshes per 50 cm. After one month of rearing, when the nets need cleaning, transfer to a net with 180 meshes per 50 cm is common practice. Juvenile fish of a length of 3—4 cm can be kept in netting with 110 meshes per 50 cm, in which nets they will stay until they have reached a length of about 10 cm. Then they are transferred to nets with diamond-patterned meshes measuring about 1 cm, and at a size of 20 cm, to net cages with meshes of 2 cm, in which they may stay till they reach the marketable size, which takes from 2—3 years.

If one starts with hatchery fry of 10 mm length, the survival to market size may be 50%, but if one begins with juvenile fishes of a length of 20 mm and more, survival is 90% and over.

B.4. Marketing red sea bream

Harvesting and transportation to the market follows virtually the same pattern as described in Chapter 6. However, Mr. Imura operates more independently than the Kita-Nada growers, selling part of his crop in Ehime Prefecture, in particular in Matsuyama. In that case he does not use a hired well-boat, but his truck, cooling his fish with ice. Consignments for Osaka and Hiroshima travel by well-boat.

C. FARMING RISKS

The most precarious phase is taken care of in the hatcheries. Not all the eggs will hatch and quite high losses may be observed in the period that the larvae have to switch over from yolk to more substantial feed to be ingested via the mouth. Deterioration of the water may lead to serious losses, but so can excessive growth of diatoms, leading to gas bubbles inside the fish larvae which makes them float at the surface, to die in 2 or 3 days. Another hazard in the fry stage is calculus formation in the bladder, a question of food deficiency.

The high survival figures given for juvenile fishes from 20 mm upwards indicate that parasites and diseases thus far do not levy a heavy toll in red sea bream farming.

D. ECONOMIC ASPECTS

D.1. Inventory items

Of the inventory items of Mr. Imura's enterprise, the wooden pier and the 15 net frames moored at it should be mentioned. Further, the freezer of 200 ton capacity, a truck, a 5-ton boat, and three smaller boats. For the net frames one needs a supply of a fair number of net cages of different mesh size. Prices for boats, net frames and netting are given in Chapter 6.

D.2. Expendable items

Purchase of fry is an important item in the budget. When one starts rearing with juvenile fishes, measuring 20 mm and over, caught by fishermen, one may count on a survival rate as high as 90%, but every individual fish will cost as much as Y 130—150. Mr. Imura purchases thousands of juvenile sea breams, sometimes more than 10 000 every year.

As in yellow-tail farming, feed is an important expense. If the price of feed goes up steeply, as was the case in the year 1974, the profitability of the enterprise may become badly affected.

D.3. Sales

Usually the red sea bream is sold at a weight of 1—1.5 kg. At that weight, the gross price the producer obtained was about Y 1 240 per kg in the period of observation. As has been said above, the consumer may pay as much as Y 3 000 per kg for specimens caught in the wild, preferred to the farmed sea breams because of their more attractive colour and better flavour. In 1973 Mr. Imura sold 5 000 four-year old sea breams with an average weight of 1.4 kg and in 1974 10 000 sea breams three-years old with an average weight of 1 kg. It should be taken into consideration, however, that only three of his 15 net cages were used for sea breams in the period of observation, the others being used for yellow-tail. The cages held then respectively 6 000 one-year old sea breams, 3 000 two-years old and 1 000 three-years old. Further, there were 110 000 one-year old yellow-tails and 20 000 of two years. The price made for yellow-tails varied from Y 400—800 per kg in the period of observation.

D.4. Personnel

Mr. Imura employs 10 persons full-time, four male and six female. This staff is primarily engaged in the handling, storage and transportation sections for feed fish and marketable fish, serving other farmers and fishers as well. Only one or two members of the staff are directly involved in fish farming operations. Mr. Matsune at Kita-Nada draws on the labour pool of the large and well-organized local fisheries co-operative in times of stress of work, but Mr. Imura has to call in most of his staff employed in other sections of his enterprise for special operations such as harvesting the marketable fish.

E. GOVERNMENT ASSISTANCE

The attitude of the Government towards farming of food fish, and the various ways in which the farmer receives assistance, is for red sea bream so much the same as for yellow-tail that no special description need be given here.

F. GENERAL IMPRESSION

The total tonnage of red sea bream farmed now is not yet very impressive, but nevertheless quite a few farmers in southern Japan, including the Seto Inland Sea area, make a living on it, often in combination with yellow-tail farming to balance the risks and to take advantage of occasional high prices.

That so many rather costly efforts are made to produce sea bream fry in hatcheries is not so much a result of requests from the farmers, fearing a shortage of natural fry in the near future, but rather from the general program envisaging the release of large numbers of juveniles of non-migrating species of fish and shellfish held in high esteem by the consumer but becoming scarcer and scarcer in the Seto Inland Sea. As far as research is concerned, there does not seem to be an agreed system of priority studies in this field, apparently because the prefectural centres have an immediate responsibility to their own region, and local political pressure is surely a powerful factor.

As has been said before, the hatchery policy is predominantly guided towards releasing juveniles in overfished waters. That farmers now buy large numbers of hatchery fry is an additional development of growing economic importance. Practical results of the releasing program have yet to be demonstrated. Allen and Shelbourne said about this, in their unpublished report: "Millions of fish and shellfish are released annually into the Inland Sea, not as part of an experimental program to assess the effect, but as an accepted remedial measure".

The future may show whether the release produced positive results, but to observe that properly, an efficient monitoring program will have to be set up.

Farming the red sea bream is certainly an interesting industry, much to the benefit of the farmers who make a living on it, and of the consumer who is willing (under the gentle pressure of tradition) to pay a very high price for the marketable product. This makes it possible to farm the red sea bream with profit, though its growth rate is much too slow to warrant remunerative farming if one were aiming at production of mere food. In fact, it is a luxury item, produced at the cost of a far greater tonnage of perfectly edible fish of other species, but it surely contributes to the well-being of the Japanese people.

G. BIBLIOGRAPHY

Allen, J.H. and Shelbourne, J.E., 1966. Unpublished report on a visit to Japan during September 1966. Fisheries Laboratory, Lowestoft.

Hushimi, T., 1968. Studies on production of young fish of the red sea bream (*Pagrus major*). Effects of a few kinds of food upon growth and survival rates of fry. Rep. Fish. Exp. Stn., Hiroshimakan, 1: 49—54.

Tamura, T., 1970. Marine aquaculture. Nat. Sci. Found. NSF, Washington, D.C., PB194, 051T.

Chapter 8

FARMING RAINBOW-TROUT AND SALMON IN THE MARINE ENVIRONMENT IN NORWAY

Example
Fish farm of Erling Osland in the Fuglesettfjord near Bjordal, Norway.
Period of observation: 1970—1971.

A. BACKGROUND

A.1. General principles

Although carnivorous fish can be grown in captivity, examples of remunerative farming of such species in the marine environment are few, at least in countries where high wages prevail. It was Norway which took the lead by initiating farming of Salmonids in salt water a few years ago, and it can already be said that farming of salmon and rainbow-trout in the marine environment is a considerable success, and also that it appears economically very promising for any countries in which the hydrographic and biological conditions are suitable.

Salmonid fishes are carnivorous and therefore belong to the third and fourth links in the food-chain. Economically this means that the animal protein they have to be fed on should not only be readily available at any time of the year, but it should be cheap too. Considering that food is only one of the items of expenditure, it should, by preference, not cost more than about 5% of the price per unit weight to be ultimately obtained for the marketable fish. Some salmonid species spend part of their life in fresh water, where they breed, and part in the sea, where they can find a wealth of food for rapid growth and fattening. This explains why such species are quite adaptable to different salinities, and though their early development has to take place in fresh water, it is possible to transfer them later, with due care, to sea water if this offers certain farming advantages. That the salmon itself may, in nature, migrate over thousands of miles need not be interpreted as bar against growing this species smoothly in confinement; the Norwegian fish farming showed that extensive migration is not a prerequisite for the salmon's well-being.

The conditions required for salmon and rainbow-trout farming in a marine environment are:

(i) Availability of an adequate supply of pure and cool fresh water for the hatchery and for the rearing of fingerlings;

(ii) sea water and fresh water, combined to fill the rearing ponds with water of the desired salinity;
(iii) sufficient shelter to avoid gale damage;
(iv) sea water kept in motion by the tides to ensure adequate flushing of the water in the rearing pens;
(v) ice-free conditions in the bodies of sea water in which the pens have to be constructed;
(vi) an adequate and regular supply of low priced food.

Such conditions are fulfilled in many sites in the coastal area of Norway between the city of Haugesund and the Vestfjord.

The farming of salmon and rainbow-trout can be subdivided into four phases: (a) the hatchery or indoor phase; (b) rearing young fish to fingerlings in troughs or tanks; (c) growing the fish to a bigger size in earth ponds filled with brackish water; (d) further growth and fattening in pens constructed in water of high salinity.

For the phases (a) and (b) one could lean on the wealth of experience gained in rearing salmonid fry in hatcheries of various descriptions, whereas phases (c) and (d) constituted a new venture, which soon led to a noteworthy success, promising development of a thriving industry in countries like Norway and Scotland, where natural conditions are suitable.

Unlike the farming of most of the species of the second link in the food-chain — rightly called semi-farming — the rearing of salmonids in the marine environment is a typical case of full-cycle farming, for not only is the production of fry carried out under controlled conditions, but also the provision of food in the later stages is fully under the control of the farmer; and adult fishes from the farm itself can be used as parent-stock, which makes the farm independent of hazardous introduction of stock from elsewhere.

A.2. Biology of the species

There is no need to describe the biology of the European salmon, *Salmo salar* L., and the rainbow-trout, *Salmo gairdneri* Richardson, in detail here. It is important to note (for fry-rearing) that an adult salmon produces 1 000—2 000 eggs per kg of body weight; the eggs are rather big (about 6 000 to a litre), so that each larva receives an appreciable quantity of reserve-food in its yolk-sac. There is, therefore, no need to feed the newly hatched larvae. From the moment the yolk-sac is resorbed, the larvae can be fed on dry food, specially produced for rearing fry of salmonids.

Salmon eggs are produced in October and November, and require about 440 day-degrees for hatching (i.e. 70—200 days, depending on the water temperature). It takes an additional 2—3 weeks to resorb the yolk-sac, after which the young fish can be grown in outdoor troughs on dry food. Some sea water can be then added, to give a salinity in the troughs of 5‰. As the juvenile fish grow they are distributed in a greater volume of water, receive food of greater particle size, and are kept in water of higher salinity. In their second summer the young fish are transferred to pens constructed in a body of tidal sea water, and only from then onwards are they fed with natural food, a ground mixture of fish and shrimp peelings.

The larger the fish and the higher the water temperature, the greater is their adaptability to salinity changes.

The biological characteristics of the rainbow-trout of special importance for farming do not differ very much from those of the salmon, although the former species is usually grown in fresh water until maturity.

The rainbow-trout spawn later than the salmon, in February—April. An adult female spawns from 1 000—5 000 eggs, according to its size, and the eggs require some 350 day-degrees for hatching. The juvenile rainbow-trout grows faster than the young salmon, and can be transferred to pure sea water after having been fed on dry food for about one year.

A.3. Geographical situation

The fish farm of Erling Osland is to be found on the western bank of the Fuglesettfjord, due north of the bridge leading to Bjordal. The Fuglesettfjord is a branch of the Sognefjord, on its left bank some 60 km inland. The Sognefjord is situated in the district Sogn og Fjordane on Norway's west coast, some 80 km north of the city of Bergen.

The exact geographical position of the fish farm is 61° 05′N and 5° 45′E.

A.4. Hydrographic pattern

The hydrography of the Fuglesettfjord very closely resembles that of the Sognefjord itself. The Sognefjord stretches west—east far into the interior. To Skjolden, at the end of the northeastern side-branch, it measures 195 km in length, and to Årdal, due east, 168 km.

Over the greater part of its length the Sognefjord is about 5 km wide and in the narrower branches, further east, its breadth is some 2 km. It shows the typical U-shape of a Norwegian fjord: rocky bluffs and a broad flat bottom covered with sediments brought in by the rivers. The Sognefjord is quite deep; more than 800 m over most of its length, but up to 1 300 m in the section where the Fuglesettfjord branches off. On its western side one finds a threshold at 150 m depth. This configuration leads to layering of the water. The topmost layer, 5—15 m thick, receives the water from the rivers and is therefore somewhat brackish, especially in the summer months when the snow on the mountains melts. This layer flows seawards. The middle layer, down to the threshold depth of 150 m, is in regular exchange with the water of the Atlantic Ocean due to the tides, with a net-flow towards the land, whereas the layers at greater depth are more stagnant, exchange being mainly

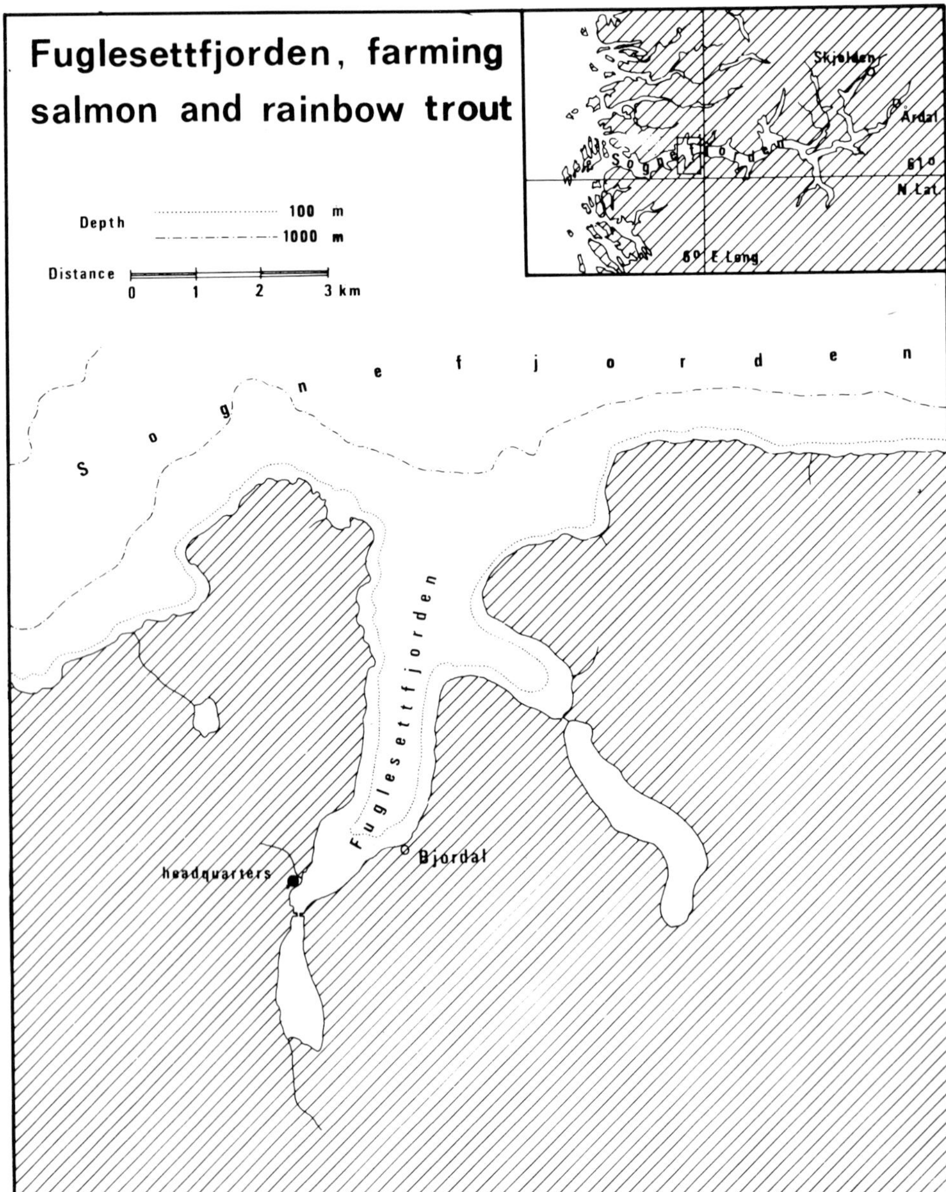

by diffusion processes. In occasional years there is a stronger influx of sea water to the middle layer.

The hydrographic pattern of the Fuglesettfjord itself has much in common with that of the Sognefjord. It is shallower, but over most of its length more than 100 m deep, which means that seasonal layering of the water masses is a normal feature here too.

Fig. 78. The Fuglesettfjord seen from the south. The fish farm is on the western bank beyond the bridge. (Photo P. Korringa.)

Tides. The Sognefjord is tidal, showing a range of 1.50 m at spring tide, and of 0.80 m at neaps. This is of paramount importance for adequate flushing of the pens of the fish farm.

Water temperature. Only the upper layer of the fjord water shows noteworthy seasonal fluctuations in water temperature. In layers over 100 m deep the temperature stays between 6.5 and 7.5° C all year round. In the central section of the Sognefjord water temperatures of the topmost layer fluctuate around 5° C in the first three months of the year, to rise slowly to a maximum of 16° C in July and August, which leads to an annual average of 9.7° C. But at a depth of about 20 m the winter temperature is 6° C and the summer maximum only 10.5° C, leading to an annual average of 8.2° C.

Salinity. In the deeper layers of the Sognefjord salinities are very steady at 34.90—35‰, but the surface layer shows a marked seasonal salinity pattern. From November to April the average influx of fresh water in the whole fjord amounts to some 40 m^3/s, to rise in May to 300 m^3/s and in June—July to 550 m^3/s, dropping again to 460 m^3/s in August, 350 m^3/s in September and 180 m^3/s in October. The influx of fresh water reaches an annual total of 8 km^3 of which 5.8 km^3 is contributed by seven rivers of some importance. The fresh water makes its way to the sea through the surface layer of the fjord, but since some 18 km^3 of sea water penetrates each year in the middle layer, the total drain-off through the top layer amounts to some 26 km^3. The total content of the Sognefjord is some 1 768 km^3, of which 525 km^3 pertains to the topmost 100 m.

All this leads to a typical salinity pattern in the topmost layer: in the winter months the salinity is high and steady over a considerable depth range, from 31‰ to 34‰, but in April layering begins to appear; in the topmost layer the salinity drops steeply to 14‰ or less. Data collected in the Fuglesettfjord show that the topmost 5 m are 14—18‰ saline from June till September, whereas at a depth of 5 m the salinity does not drop lower than to 15‰ over that time, and at a depth of 10 m not lower than to 29‰.

This salinity pattern, with reduced values in the top layer in the summer months, accompanied by rather high water temperatures in the upper layer, makes it possible to ensure a very gradual salinity transition when the fish are transferred from the earth ponds to the pens in the open fjord in midsummer.

A.5. Legal aspects

The fish-farm of Mr. E. Osland is a private enterprise. He is assisted in all operations by his two sons-in-law, Mr. O. Ingebrigtse and Mr. A. Brekke, both of whom have received training as fishery biologists at the University of Bergen, which makes them highly qualified assistants. Mr. Osland has no need to work with hired hands. It would, moreover, be rather difficult to find reliable hands in this sparsely populated part of the country, and costly to provide them with adequate housing.

To the farmers who own the land facing the Fuglesettfjord, Mr. Osland has to pay an annual fee of NKr. 10 per running metre of coastline for the sites where he constructs his pens in the fjord.

Theft has never occurred here, so surveillance is deemed unnecessary.

B. SEQUENCE OF OPERATIONS

B.1. The hatchery

Close to his residence Mr. Osland has erected a concrete structure 7.50 × 5.50 × 2.10 m. In this hatchery building are placed 10 fibreglass troughs, the sidewalls tapering slightly downwards (brand LM, Danish made), measuring 365 × 42 × 17 cm, plus one of half that length. At one end of the trough there is an inlet, and at the other end an overflow outlet pipe. Just inside the latter there is an inclined wooden frame with a screen of perforated aluminium. In these troughs with running water are placed tapering square trays measuring 42 × 42 × 15 cm made of the same material, seven in the long troughs, four in the shorter troughs, a protruding rim, fitting on the sidewalls of the troughs, keeping them a few centimetres off the bottom. The entire bottom and a 37.5 × 8 cm section of the hindmost wall of the trays consists of 1 mm-thick aluminium sheet with perforations of 1.5 mm diameter placed in alternate rows (produced by the Norsk Perforerings Fabrik, Vikersund). The hindmost wall of each tray carries at its under end a solid glass fibre strip 3 cm high, which blocks the water current along the bottom of the trough, whereas the sidewalls of trays and trough fit together, leaving no free space for the water. The water entering the troughs through the inlet is thereby forced to follow its way from the bottom of the first tray upwards, to leave through its hindmost wall, then diving down again to find its way through the bottom of the second tray, etc. This is called the Californian system. Two or three litres of fertilized eggs of salmon (October—November) or of rainbow-trout (February—April) are placed in each tray and some 8—10 l/min of water, to be used once only, should run through during the whole period of incubation, i.e. 80—100 l/min for the whole battery. This gives the hatchery a capacity of some 500 000 eggs for both species, with a slight overlap at the last phase of salmon hatching to the onset of trout egg incubation in early spring.

The quality of the water which passes through the trays with developing eggs is of the utmost importance. A long pipeline guides the pure, clear water from a brooklet up in the mountains to the hatchery, but before being led into the troughs it has to pass a big tank filled with marble grit (pieces of 2—3 cm). The water is forced to follow a long path through this filter in order to eliminate all particulate matter, to raise the pH somewhat, and to adjust its gas content. This latter is necessary since the pressure in the pipeline leads to oversaturation with air, which would lead to air bladder disease in the juvenile fish if not brought down to the level normal for atmospheric pressure.

Salmon eggs require 450 day-degrees for hatching and rainbow-trout eggs 350, but the latter should preferably not be exposed to very low water temperatures. Especially during the first days after fertilization, the water temperature should be kept at about 4—5°C for rainbow-trout eggs, and later, if possible, not lower than 3°C. Since the water from the brooklet in February—March is 0.5—1°C, it must be heated. A heat exchanger has been installed in the hatchery, working on electricity. Since new water flows in continuously, heating is quite an expensive item on the hatchery budget. Moreover, the electricity supply is not fully reliable in this remote part of the country, so the heater often works intermittently.

After hatching, the larvae are kept for 2—3 weeks in the trays, until the yolk-sac is fully resorbed.

B.2. The outdoor rearing troughs for the young fry

Close to the hatchery are a series of 16 troughs and a battery of nine round fibre-glass tanks, destined for receiving the young fry from the hatchery, the former generally used for trout, the latter for salmon.

The troughs are of the so-called Danish type, 7 m long by 86 cm wide in the clear, and 57 cm deep, made of concrete some 9 cm thick. The troughs draw their water from a long reservoir placed crosswise at a somewhat higher level, receiving it in its turn from the mixing pond. The water is led into the upper end of each trough, either through an overflow pipe or through a rub-

Fig. 79. General view of Mr. Osland's fish farm. In the foreground the mixing pond, with inflow of sea water, from which, water of 5‰ salinity is led into the outdoor rearing troughs (left). In the background the earth ponds and the freezing house, annex, storage room and workshop. (Photo P. Korringa.)

ber tube led to the bottom layer of the trough. This tube is especially used when salmon fry are kept in the troughs, for they prefer a good flow of water along the bottom. Each trough should receive water at the rate of 40 l/min. Near the other end of the trough is placed an inclined wooden frame carrying an aluminium sheet with a number of slits of 20 × 2 mm, 4 mm apart. The used water leaves the trough through a wide plastic tube inserted near the bottom and kept in the desired position by placing a piece of board under it. It overflows into a gutter at a height slightly below the rim of the trough. There is a second hole in the terminal wall of the trough, made at bottom level and normally plugged up. It is only opened to clean the trough, letting the water level drop quickly, causing a good current to take away the deposits.

The concrete walls of the trough should be as smooth as possible; they are painted with "Interrasing", a type of anti-fouling paint also used for boats, which inhibits algal growth. It is advisable to give the troughs a new coating of paint every year, but after thorough drying one should flush the troughs for days in succession and, to avoid all risks, next place a few small fish in it for some time to observe any ill effects of poisonous components of the paint.

The water for feeding the troughs comes from the same source as that used in the hatchery. It is led into a mixing pond of some 5 × 10 m × 50 cm, the bottom of which is strewn with marble grit. Sea water is pumped in continuously from the fjord, carefully metered to obtain a salinity of 5‰.

B.3. Maintenance of the juvenile fish in the troughs

Thirty thousand juvenile rainbow-trout are transferred to each trough when their yolk-sac has been resorbed. Now they should be fed, using dry pellets especially produced for rearing salmonid fry (e.g. of the Swedish brand EWOS). These pellets, containing all the necessary nutrients, are available in a whole series of particle sizes, the smallest almost dust-like in appearance, the next like small sand-grains, etc. The fry is fed at frequent intervals, virtually every 30 min, from 8 a.m. till 7 or 8 p.m., later at greater intervals. One gives as much dry food as is taken eagerly, which means 5—10% of the body weight per day. For correct dosing of the food for the small fry one uses a plastic medicine bottle which produces exactly 10 cm^3 of fine grained pellets at a squeeze. To assess the correct dose for trout of different size use can be made of Table II (p. 194).

The troughs should be kept as clean as possible by frequently eliminating all excess food and droppings. To do this, the water level in the trough is lowered by opening the bottom drain, leaving a layer of about 5 cm of water. A rubber hose connected with the reservoir is then used to drive any dirt or deposits towards the outlets. This is initially done every other day, later once per week, and takes 15—20 min per trough.

TABLE II

Daily quantity of dry "TESS" feed (kg per 100 kg of rainbow-trout)

Trout weight (g) Size (cm)	0.2 2.5	0.2—1.5 2.5—5	1.5—5 5—7.5	5—12 7.5—10	12—23 10—12.5	23—40 12.5—15	40—62 15—17.5	62—90 17.5—20	90—130 20—22.5	130—180 22.5—25	180 25
Water temperature (°C)											
5	3.8	3.0	2.5	1.7	1.4	1.1	0.9	0.8	0.7	0.6	0.5
6	4.0	3.2	2.6	1.8	1.5	1.3	1.0	0.9	0.8	0.7	0.6
7	4.5	3.5	2.8	1.9	1.7	1.4	1.2	1.0	0.9	0.8	0.7
8	4.8	3.8	3.0	2.4	1.8	1.5	1.3	1.1	0.9	0.8	0.7
9	5.0	4.3	3.4	2.7	2.0	1.7	1.4	1.2	1.0	0.9	0.8
10	5.5	4.5	3.6	2.8	2.1	1.8	1.5	1.3	1.1	1.0	0.9
11	6.0	4.8	3.9	3.0	2.3	1.9	1.6	1.4	1.2	1.0	0.9
12	6.5	5.2	4.2	3.2	2.5	2.0	1.7	1.5	1.3	1.1	1.0
13	7.0	5.5	4.5	3.5	2.6	2.1	1.8	1.6	1.4	1.2	1.1
14	7.5	6.0	5.0	3.7	2.8	2.3	1.9	1.7	1.5	1.3	1.2
15	8.0	6.5	5.5	4.0	3.0	2.5	2.0	1.8	1.6	1.4	1.3
16	8.5	7.0	5.8	4.8	3.5	2.6	2.1	1.9	1.7	1.5	1.4
17	9.0	7.8	6.3	5.0	4.0	2.7	2.2	1.9	1.7	1.6	1.5
18	9.7	8.5	6.8	5.5	4.4	2.8	2.3	2.0	1.8	1.7	1.6
19	10.4	9.2	7.1	6.0	4.6	3.2	2.4	2.1	1.9	1.8	1.7
20	11.0	10.0	7.0	6.3	5.0	3.5	2.5	2.2	2.0	1.9	1.8

As a rule the rainbow-trout fry grows well in the troughs because water temperatures are already somewhat higher at the time they are transferred to the outdoor troughs. After about 4 weeks they need more space, so the population of a trough is distributed over two units. When the fry have stayed for about 6 weeks in the troughs they are big enough to be transferred to the large earth ponds. Usually the losses are under 10%, but occasionally (e.g. in 1971) losses rise to 30 or 40%.

B.4. The round fibre-glass tanks for rearing salmon fry

Close to the series of concrete troughs is a battery of nine round fibre glass tanks, about 1.5 m in diameter and of 38 cm in depth. In the middle of the bottom there is a square sheet of perforated aluminium, which leads to a plastic outlet tube. By adjusting the height of the far end of the tube the water level in the tank can be regulated. The water is squirted in from the mixing pond (salinity 5‰) tangentially, producing a constant circular motion, which makes the deposit collect on the centrally-placed screen. By lowering the overflow outlet a more rapid stream of water is produced through the screen, which takes along all excess food and droppings. A long-shafted brush serves to clean the screen whenever necessary, and a small dip-net is used to take out any dead specimens.

Ten thousand salmon fry are put in each tank and adequate flushing is ensured, which for salmon fry is more intensive than for the rainbow-trout troughs.

After about 2 months the growing juvenile salmon require distribution over two tanks. The salmon grow more slowly than the rainbow-trout, partly because they are brought into the tanks or troughs earlier in the year, when water temperatures are still very low. They are kept in tanks or troughs until October.

B.5. Transfer of the fingerlings to earth ponds

The young rainbow-trout is transferred to large earth ponds after about 6 weeks in the trough, in August or September of the year they were born, but the young salmon which hatched earlier in the year are kept in the tanks till October or November. Close to the shore of the fjord there is a group of four ponds measuring 25 × 5 m × 1.20 m. This complex has earthen walls and bottom, but the partitions creating the four sections are made of wood. Fresh water from the mountain brook is led into the ponds, but here again, continously admixed with sea water pumped up from the fjord. When the fish are transferred from the trough to the ponds the water in the ponds is of 5‰ salinity, but gradually the quantity of sea water is increased so that the ponds show, after about one year, a salinity of 15‰. The ponds should be flushed continuously, but often the supply of fresh water is too limited here, especially in the winter, and hatchery and rearing troughs have priority over the ponds. Usually all the young fish from the troughs are brought with a dip-net into one pond section which can then get a good flushing, and later they can be distributed over a greater pond acreage. The young rainbow-trout and the juvenile salmon transferred to the ponds continue to be fed with dry pellets, either of the Swedish brand EWOS or a special trout food "Tess", produced by T. Skretting a/s, Stavanger. One should count on feeding with pellets for a whole year after resorption of the yolk-sac, continuing to offer the young fish 5—10% of their body weight per day; more at higher water temperatures than at lower. When the ponds are properly maintained and the food is dosed carefully, there is no need to take measures to keep the oxygen content of the water at the desired level.

When the fishes have grown for some time and should be distributed over more pond sections, they are taken out with a seine-net; not an easy procedure since the level of the pond is too close to the fjord level to allow emptying of the pond by gravity. Sometimes electric pumps must be used to bring the pond level down sufficiently.

Apart from the four large ponds there are two spare ones, which late in the year are used to keep a well-selected stock of adult fish for spawning. The salinity in these ponds is then 15‰. This is considered best for producing high-quality eggs and sperm; the salinity in the pens in the open fjord rises to 30‰ late in the year.

B.6. The pens in the open fjord

The first pen was constructed here in the year 1964; in 1971 there were four in use, three for rainbow-trout and one for salmon. Each enclosure is 600—700 m^2 and is constructed at the sloping edge of the fjord, down to a depth of 6.5 to 7 m. The shape of the pen is trapezoidal; two concrete walls protruding from the solid grounds serve to attach the netting. On the tapering sides, which are about 20 m long, and along the far end of the trapezium, long square poles of impregnated wood have been driven into the ground, 5 m apart. This has been done using a hired pile-driver.

These rows of poles serve to support the netting. To keep injuries to the delicate fish down to a sheer minimum, a knotless nylon netting has been selected; it is impregnated with a copper compound and has square meshes, 1 cm for trout, 2 cm for the larger salmon. It is made by IFA in Bergen. Although one counts on a life of at least 10 years for the netting, a single hole might lead to the loss of all fish in the pen. Therefore, Mr. Osland decided to place a double wall of netting, one attached to the outside of the poles, the other to the inside. Just above the bottom, the poles carry metal rings through which a strong rope is strung to keep the nets close to the poles, and an extra strip of netting, 100 cm broad, carrying lead in the seam, lies horizontally on the bottom of the pen, to prevent escape under the netting. Supporting ropes attached to the poles keep the netting nicely stretched. Every second or third year the netting is taken up to be repaired and re-impregnated.

B.7. Maintenance of the fish in the pens

A pen is stocked with 200 000 up to 250 000 yearling rainbow trouts, or with 100 000 young salmons. This is usually done in May or June. To keep stress down to a minimum, one tries to keep the salinity in the pond and in the pens about the same at the time of transfer, which is fairly easy here since the surface layers of the fjord water show a reduced salinity in the sum-

Fig. 80. A double wall of netting prevents escape of fish from the pens in the fjord. (Photo P. Korringa.)

mer season, and the ratio of brook water to fjord water supplied to the ponds can be regulated.

The fish should be at least 10 cm long when transferred to the pens. They will receive from then on another type of food, consisting of fresh fish. This fish food is selected from the species which are readily available, either fresh or deep-frozen: mainly sprat (*Clupea sprattus*), herring (*Clupea harengus*), coalfish (*Gadus virens*), Norway-pout (*Gadus esmarkii*), heads of dogfish (*Squalus acanthias*), etc. Capelin (*Mallotus villosis*) would be preferred for salmon, but is caught only as far away as Finmark and therefore is not readily available here. The fish is usually bought deep-frozen, but should be quite fresh when frozen. After thawing out (the frozen blocks are placed in running sea water), the fish is minced into morsels of about 1.5 cm diameter for feeding the larger fish, and passed through a second mincer to produce morsels of 0.5 cm when smaller fish have to be fed. Sharp bones, such as those found in coalfish, should be avoided whenever possible for they can cause injuries in the salmon's gullet. A special ingredient, frequently admixed in the food, are shrimp peelings, available in Norway in ports where the deep-sea prawn *Pandalus borealis* is landed. These contain a fair amount of carotenoids, which give the flesh of both salmon and rainbow-trout an attractive red colour. Experience has taught that cod roe is not a full-value food item for salmonid fishes.

The quantity of wet food to be offered is about 5% of the wet body weight of the fishes per day, more at high water temperature than at low. The fish in the pens are fed at frequent intervals from 9.00 a.m. till 5.00 p.m. by spreading the ground fish by hand from a wheel-barrow placed on the catwalk above the pen. The fish are not fed on Sundays.

A considerable advantage of the pens being in tidal sea water is that a far greater number of fish can be kept in such a pen than in a fresh water pen, with practically stagnant water, of similar dimensions. The currents take away all excess food (including particles too small to be grabbed) and liquid components, together with the fish droppings. The water in the pens will therefore never foul, and the oxygen level is always suitable.

It takes rainbow-trout, placed in the pens at a length of 10—15 cm, about 1.5 years to attain a weight of 1—2 kg. They are then roughly 2.5 years old. This means that they grow much better in the pens in the marine environment than in the usual fresh water ponds. Salmons weigh from 1—4 kg at an age of 3.5 years and 2—8 kg when they are 4.5 years old. They often reach sexual maturity rather early, at an age of 3.5 years, which reduces their market quality.

B.8. Harvesting the marketable fish

The marketable fish is taken from the pens with a seine-net, carefully stripped and washed, and then packed in plastic boxes which can hold either

18 or 30 kg. After ice has been added they are shipped to the market. Since deep-freezing facilities are available on the spot, salmon and trout can also be stored that way. When individually packed in plastic packets filled with sea water, they give a deep-frozen product of excellent quality. To get the best dish of salmon or sea trout one should keep them for two days in a cool place before offering them to the cook.

C. FARMING RISKS

When all the safety measures described above have duly been taken into account, the risks in farming rainbow-trout and salmon in pens in the open fjord are apparently much smaller than in fresh water farming. At least some of the parasites and diseases one has to cope with in fresh water ponds seem to be absent at higher salinities. Moreover, growth is faster in sea water pens, which may make the fish more resistant to various hazards.

Fish farms often suffer from explosive development of parasites and diseases after introduction of parent stock or fingerlings from elsewhere. One of the advantages of the type of fish farming developed here is that parent stock reared in one's own pens can be used, both for salmon and rainbow-trout, and all the fingerlings needed are bred on the spot. This is certainly an important feature in reducing farming risks.

There are only a few predators to be kept under control. Herons and kingfishers do not occur in this area, but terns and some species of sea gulls try to feast on the young fishes. The small fry in the troughs and tanks are not waylayed by these birds, presumably because of their proximity to the residence, but both the ponds and the pens have to be protected against their attacks. A simple and effective measure to reduce predation by birds is stringing series of nylon wires over ponds and pens.

D. ECONOMIC ASPECTS

D.1. Rents

For every running metre of shore line in use for fish farming a fee of NKr. 10 per year has to be paid to the owner of the land.

D.2. Inventory items

By far the most expensive inventory item is no doubt the freezing house, a two-storey wooden building measuring 15 $\times$ 17 m. It contains a 7 $\times$ 7 $\times$ 5 m deep-freezing section (—30° C), further a cool storage (0° C), rooms for thawing and grinding fish, storage space for various materials, and a workshop.

Two motorized mincing machines are installed in the shed. Other items of importance are a 1.5 ton fork truck with electric battery, a 60 h.p. tractor (brand Power Major) to transport among other things the wooden boxes with ground fish, and a fibre-glass row boat. Further, there are three electric pumps in use, of 2, 3 and 5 h.p. respectively.

The double netting of the four pens should also be taken into account (estimated life 10 years) and further, the inventory (troughs, trays, tanks, heat-exchanger) of the hatchery and for the open-air rearing of fingerlings.

D.3. Expendable items

Food is an ever-recurring item on the budget. The pellets are purchased by the 1 000 kg, the small particle-size pellets costing NKr. 2.50—3.00 per kg, the coarser pellets NKr. 1.50 per kg. One has to feed 5—10% of the fish body weight per day, which means that the daily expenses for feeding increase with the size of the individual fishes, and are of little importance in the first few months of rearing. For both salmon and rainbow-trout one should count on a full year of pellet feeding after resorption of the yolk-sac. To feed the fish in the pond in 1971 (four pens stocked), 1 000 to 1 800 kg of fresh fish were needed per day. A ship loaded with 20—40 metric tons of deep-frozen fish comes to the plant to unload its cargo about once per fortnight, which requires 8—10 h sailing time. The deep-frozen fish costs from NKr. 0.30—0.70 per kg.

Other expendable items are the electricity required for the freezing house, the heat exchanger in the hatchery and for operating pumps and mincing machines, and fuel for the tractor and fork-lift truck. Ice and packing material have to be bought in the appropriate season.

D.4. Sales

The production of rainbow-trout by the Osland farm fluctuated in recent years between 20 and 40 metric tons per year, for a wholesale price of NKr. 12—15 per kg. The salmon production is still on the increase since the difficulty of obtaining suitable parent stock has only recently been overcome; they now come from the farm itself. The production in 1971 was about 10 metric tons, for which a price of NKr. 15—40/kg is made. It should be taken into account that the customer is willing to pay up to NKr. 75/kg for top quality salmon in the season that they cannot be obtained from the free fishery, and that in Norway a retail price of NKr 35/kg is now paid for top quality red-fleshed rainbow-trout.

D.5. Personnel

There are no hired hands on the fish farm of Erling Osland. He and his two sons-in-law do all the work, which means that three families have to make a living from the farm.

D.6. *Further development*

The Osland fish farm has not been established for long and is still in development. One of the plans is to construct three more pens for salmon so that the farm will operate with three pens for rainbow-trout and four for salmon, keeping all salmon year-classes in separate pens.

Another plan is to transfer the hatchery to Matre, at the head of the Masfjord, a branch of the Fensfjord. In Matre (60°50′N, 5°39′E) a power plant has been constructed. This means that one can use its fresh cooling water. This would not only drastically reduce the costs for electric heating of the hatchery water, so that the water could be kept exactly at the optimum temperature throughout the whole period of incubation, but there would be the possibility of constructing rearing troughs and fish ponds in the vicinity of the Matre power plant, thus overcoming the recurring difficulty of seasonal shortage of fresh water in the present ponds. There is also more room for improving the farming operations. Thawing the blocks of frozen fish, mincing the fish, transporting the mixture to the pens and spreading it in the pens still requires a considerable amount of manual labour. Mechanization is certainly possible to a large extent. The same holds good for the distribution of dry food in the troughs and the ponds. Various modern fish farms have introduced automatic dosing of pellets. If fish like coal-fish has to be used, it would be good to introduce a filletting machine to get rid of the hazardous coarse bones.

Since not only the available volume of fresh water, but also the amount of manual work are now the bottle necks in this farm, mechanization and automation are steps which deserve serious consideration if the production is to be increased.

E. GOVERNMENT ASSISTANCE

The Norwegian Government follows the developments of farming of rainbow-trout and salmons in pens in the marine environment with a keen interest, and does not hesitate to assist the farmers with research and advice to further their promising venture.

The hatchery at the Matre power plant is one of these new developments, and it has been decided that a laboratory will be established there to facilitate research on the primary phases of farming salmonids. One of the problems which require a scientific approach is the provision of food of adequate composition in optimal quantities, and with a frequency leading to the best possible conversion factor. Already, now, one has the impression that the rainbow-trout in the pens will give a conversion factor of 5, which means that one can expect to produce 1 kg of rainbow-trout for every 5 kg of fish fed to them. Studies on optimal spreading of the feeding over 24 h and over 1 week could lead to results of economic importance. The faster the growth, the shorter the fish have to be kept in the expensive facilities. If light should turn out to be a limiting factor in active feeding and growth, artificial aids could be considered. Another research project could focus on the reason why the salmon in the pens tend to develop their gonads somewhat earlier than in nature. It must be possible in principle to develop a system of hormone injection to regulate the maturation process.

The fish farmers in Norway are not in a position to carry out such research themselves, but they can rest assured of the keen interest and close collaboration of the scientists of the Norwegian Fisheries Institute at Bergen (Fiskeridirektoratets Havforskninginstitutt).

An important aspect from the beginning of farming salmonids in the marine environment was that, according to Norwegian law, salmon and trout did not belong to the realm of sea fisheries, but rather to agriculture. This made it difficult to start a salmonid farm in the coastal area. But since the Norwegian authorities understood how economically interesting such ventures could be, these legal difficulties were ironed out.

F. GENERAL IMPRESSION

Farming carnivorous species of fish or invertebrates, i.e. beyond the second link of the food chain, seems, from an economic point of view, to be a hazardous enterprise. Animal food is costly, especially when it has to be very fresh or very freshly deep-frozen, or when the farmed species is exacting with regard to the type of food it needs. Even when a conversion factor well under 10 can be realized, one cannot afford to pay much more than 5% of the price per unit weight to be made from the end-product for the food. Usually one has to make quite a considerable investment to start a fish farm, and where labour is expensive, unless there is a high degree of mechanization and automation, this item weighs heavily on the budget. This means that the farm should be large enough to make this extra investment pay for itself.

In Europe, with its high wages, farming of carnivorous species seems at first sight hardly feasible. The more surprising it is then, to see that the salmonid venture started in Norway soon led to considerable success and could become, in due course, one of Norway's finest industries. Natural conditions are very conducive here: fresh water of excellent quality used in the hatcheries, with power plants locally producing cooling water which can be used to advantage in the hatcheries for the early phases of rearing; a strongly indented coastline offering many well sheltered sites where fish pens can be constructed safely; the Gulfstream water giving a mild climate, with ice-free conditions over a long stretch of coastline and rather mild water temperatures in winter; the tide which ensures adequate flushing of the fish pens; rich fishing grounds yielding large quantities of industrial fish suitable as food for salmonids at a modest price.

Farming beyond the second link in the food chain can only be remunerative for luxury products, making a high price on the market, never for sea food at modest price to feed the masses. The salmon and rainbow-trout farmed in Norway in the marine environment are such luxury products, for which the market shows a great and still increasing demand. The discriminating epicurean recognizes the superior quality of these red-fleshed fish and will be glad to pay a fair price in all parts of the season.

Farming carnivorous species will be economically impossible where large quantities of similar quality are landed by commercial fisheries for a price considerably lower than must be charged for the farmed product. For salmon and rainbow-trout this drawback does not exist: the demand in Europe is much greater than the fishery can ever yield, and the more so since the farmer can offer these fish at any time of the year. Already, now, only a few years after starting salmonid farming in the sea, a tonnage of some importance is obtained — one which may equal the fishery catches in the near future.

It is remarkable that a fish like salmon, performing extensive migrations under natural conditions, seems to be fully at ease in a small pen, provided it is given sufficient food of high quality, and that it may even grow faster under such conditions.

From a general fisheries point of view, use of industrial fish like capelin and Norway-pout for producing the highly-prized salmon with a conversion factor of 5 or 6 is certainly a favourable development. The bulk of the industrial fish finds its way via fishmeal to chickens and hogs, and their price per kg is very much lower than that for salmon and rainbow-trout.

If inventory items like cold storage plants, hatcheries and rearing pens, and also the pellet food to be offered in the first year, do not weigh too heavily on the prices of the end-product, it could very well be that transferring industrial species of fish, not suitable for direct human consumption, into salmon and rainbow-trout would turn out to be the most economical use of this natural resource.

It should be mentioned here that the farm of Mr. E. Osland, described above, is only one example and that there are already several farms in operation, both in Norway and in Scotland. The general pattern is similar to that described above, but the pens may be of different construction. Interesting are the anchored floating pens, consisting of chambers of knotless nylon netting, supported by floats of expanded plastic (Fig. 81). The advantage is that the bottom of such chambers also consists of netting so that food-remains and droppings disappear immediately without getting a change to foul and to

Fig. 81. A floating fish pen at Laksåvik. (Photo O. Ingebrigtse.)

affect the oxygen content of the water. A farm of such a type is to be found at Laksavik, at 62°30′N and 8°40′E.

Farming of salmon and rainbow-trout in the marine environment seems to have a bright future in countries endowed with such natural conditions as Norway, so close to the exacting market of the industrialized section of Europe, where the once-flourishing salmon fishery has dwindled to insignificance, if not already completely disappeared, as is the case in the River Rhine, once the richest salmon river in Europe.

G. BIBLIOGRAPHY

Milne, P.H., 1972. Fish and Shellfish Farming in Coastal Waters. Fishing News Ltd., London, 208 pp.

Rasmussen, C.J., 1967. Håndbog i ørredopdraet. Rhodos, København. International Science Publishers, Copenhagen, 242 pp.

Sedgewick, S.D., 1970. Rainbow trout farming in Scotland and farming trout in salt water. Scott. Agric., 49(4): 180—185.

Sedgewick, S.D., 1971. A sea farm for salmon off Norway. The Field, 238: 662—663.

Vik, K.O., 1963. Fish cultivation. Salmon and Trout Magazine, 169: 203—208.

INDEX